AN EXPERIMENTAL APPROACH TO CDMA
AND INTERFERENCE MITIGATION

T0075937

An Experimental Approach to CDMA and Interference Mitigation

From System Architecture to Hardware Testing through VLSI Design

by

Luca Fanucci
Italian National Resourch Council,
Italy

Filippo Giannetti
University of Pisa,
Italy

Marco Luise
University of Pisa,
Italy

and

Massimo Rovini
European Space Research and Technology Centre, ESA/ESTEC,
Directorate of Technical and Operating Support, Communication System
Section TOS-ETC; Keplerlaan 1, 2200 AG Noordwijk ZH, The Netherlands.
t: +31 71 5656156, e: Massimo.Rovini@esa.int

KLUWER ACADEMIC PUBLISHERS
BOSTON / DORDRECHT / LONDON

A C.I.P. Catalogue record for this book is available from the Library of Congress.

ISBN 978-1-4419-5421-3

Published by Kluwer Academic Publishers,
P.O. Box 17, 3300 AA Dordrecht, The Netherlands.

Sold and distributed in North, Central and South America
by Kluwer Academic Publishers,
101 Philip Drive, Norwell, MA 02061, U.S.A.

In all other countries, sold and distributed
by Kluwer Academic Publishers,
P.O. Box 322, 3300 AH Dordrecht, The Netherlands.

Printed on acid-free paper

All Rights Reserved
© 2004 Kluwer Academic Publishers, Boston
Softcover reprint of the hardcover 1st edition 2004
No part of this work may be reproduced, stored in a retrieval system, or transmitted
in any form or by any means, electronic, mechanical, photocopying, microfilming,
recording
or otherwise, without written permission from the Publisher, with the exception
of any material supplied specifically for the purpose of being entered
and executed on a computer system, for exclusive use by the purchaser of the work.

Contents

Authors

Luca Fanucci was born in Montecatini Terme, Italy, in 1965. He received the Doctor Engineer (summa cum laude) and the Research Doctor degrees, both in electronic engineering, from the University of Pisa, Pisa, Italy, in 1992 and 1996, respectively. From 1992 to 1996, he was with the European Space Agency's Research and Technology Center, Noordwijk, The Netherlands, where he was involved in several activities in the field of VLSI for digital communications. He is currently a Research Scientist of the Italian National Research Council in Pisa. Since 2000, he has been an Assistant Professor of Microelectronics at the University of Pisa, Italy. His main interests are in the areas of System-on-Chip design, low power systems, VLSI architectures for real-time image and signal processing, and applications of VLSI technology to digital and RF communication systems.

Filippo Giannetti was born in Pontedera, Italy, on September 16, 1964. He received the Doctor Engineer (cum laude) and the Research Doctor degrees in electronic engineering from the University of Pisa, Italy, in 1989 and from theUniversity of Padova, Italy, in 1993, respectively. In 1988/89, he spent a research period at TELETTRA (now ALCATEL), in Vimercate, Milan, Italy, working on error correcting-codes for SONET/SDH radio modems. In 1992 he spent a research period at the European Space Agency Research and Technology Centre (ESA/ESTEC), Noordwijk, The Netherlands, where he was engaged in several activities in the field of digital satellite communications. From 1993 to 1998 he has been a Research Scientist at the Department of Information Engineering of the University of Pisa, where he is currently Associate Professor of Telecommunications. His main research interests are in mobile and satellite communications, synchronization and spread-spectrum systems.

Marco Luise is a Full Professor of Telecommunications at the University of Pisa, Italy. He was born in Livorno, Italy, in 1960 and received his MD and PhD degrees in Electronic Engineering from the University of Pisa, Italy. In the past, he was a Research Fellow of the European Space Agency (ESA) at the European Space Research and Technology Centre (ESTEC), Noordwijk, The Netherlands, and a Research Scientist of CNR, the Italian National Research Council, at the Centro Studio Metodi Dispositivi Radiotrasmissioni (CSMDR), Pisa. Prof. Luise co-chaired four editions of the Tyrrhenian International Workshop on Digital Communications, and in 1998 was the General Chairman of the URSI Symposium ISSSE'98. He's been the Technical Co-Chairman of the 7th International Workshop on Digital Signal Processing Techniques for Space Communications and of the Conference European Wireless 2002. A Senior Member of the IEEE, he served as Editor for Synchronization of the IEEE Transactions on Communications, and is currently Editor for Communications Theory of the European Transactions on Telecommunications. His main research interests lie in the broad area of wireless communications, with particular emphasis on CDMA systems and satellite communications.

Massimo Rovini was born in Pisa, Italy, in 1974. He received his MD (summa cum laude) and PhD degrees in Electronic Engineering from the University of Pisa, Italy, in 1999 and 2003 respectively. Since 2002 he has been research fellow of the European Space Agency (ESA) at the European Space Research & Technology Centre (ESTEC), Noordwijk, The Netherlands, by the Communication Systems section of the Technical and Operational Support directorate. His interests lie in the broad area of VLSI architectures for real-time digital communication systems, hardware implementation and testing issues. Particularly, he has been working with CDMA systems and iterative decoding techniques of advanced forward error correction schemes.

Acknowledgements

Many people contributed to the success of the MUSIC project whose development gave us the cue for writing this book. The authors wish to express their own sincere gratitude to Edoardo Amodei, Barbara Begliuomini, Federico Colucci, Riccardo Grasso, Nicola Irato, Edoardo Letta, Michele Morelli, Patricia Nugent and Pierangelo Terreni of Pisa University, to Marco Bocchiola, Giuseppe Buono, Andrea Colecchia, Gianmarino Colleoni, Alessandro Cremonesi, Fabio Epifano, Rinaldo Poluzzi, Luca Ponte, Pio Quarticelli, Nadia Serina of STMicroelectronics, and to many more that we cannot explicitly mention here. Special thanks and a kiss go to Alessandra, Angela, and Silvia for putting up with us (not with Massimo, actually) during the final rush-outs and sleepless nights of the project first, and of the writing of the book later.

Foreword

My first touch with Code Division Multiple Access (CDMA) was during my early days at the European Space Agency (ESA) when I was involved with the development of an accurate geostationary satellite tracking system exploiting Direct Sequence CDMA. I distinctly recall the surprise to hear from my supervisor that *"the spread spectrum technique allows transmitting signals below the thermal noise floor"*. The statement was intriguing enough for me to enthusiastically accept working on the subject. I immediately fell in love with CDMA systems, as they soon revealed (both to my dismal and to my pleasure) being complex enough to keep me busy for more than a decade.

Shortly after moving to the ESA's main R&D establishment in the Netherlands, I started to regard CDMA as a potential candidate for satellite fixed and mobile communication networks. It was a pioneering and exciting time, when CDMA was still confined to military, professional and navigation applications. At ESA we developed preliminary architectures of CDMA systems featuring band limited signals, and free of self noise interference trough a simple yet efficient approach based on tight code epoch synchronization. Concurrently, we also started the earliest CDMA digital satellite modems development. The laboratory experiments unveiling the ups and downs of (quasi-)orthogonal CDMA interference where shortly after followed by the first satellite tests.

At that time a small US-based company named *Qualcomm* was moving the first steps in making CDMA technology for terrestrial cellular telephony truly commercial. And the fact that the co-founders of this small company were Dr. A.J. Viterbi and Dr. I.J. Jacobs convinced the management of ESA to financially support our modest R&D effort. While the 'religious' battle between the TDMA and CDMA terrestrial armies was taking momentum, in our little corner we went on studying, understanding, experimenting, and improving on CDMA technologies.

I had then the pleasure to closely follow the development of ESA's first mobile and fixed CDMA satellite networks while witnessing the commercial deployment of the first terrestrial CDMA networks (IS-95), and directly participating to the early tests with the Globalstar satellite mobile telephony system during my stay at Qualcomm in '96–'97. Since then CDMA technology started becoming the subject of industry courses, University lectures, and was often appearing on the front page of non-technical newspapers and magazines.

The final battle corresponded to the selection of CDMA in several flavors as the air interface for the 3rd Generation (3G) of personal communication systems: Universal Mobile Telecommunication Systems (UMTS) in Europe and Japan, and cdma2000 in the Americas. During the early days I also convinced my friend and former ESA colleagues Marco Luise and Filippo Giannetti, shortly followed by Luca Fanucci, to join the excitement and the frustrations of the satellite CDMA camp, and this was maybe the initial seed that later bloomed into this book.

While 'classical' CDMA technologies where getting commercially deployed, a truly remarkable investigation effort was taking place in the academic world about the issue of Multi User Detection (MUD) and Interference Mitigation (IM). MUD issues attracted the interest of hundreds of researchers around the world despite an initial skepticism about its effectiveness. With the authors of the book I was also 'contaminated' by the idea to develop more advanced CDMA detectors which can autonomously remove the CDMA self noise. But browsing hundreds of papers on the subject, we were still missing inspiration for some technique which can be readily implemented in the user terminal of a satellite network.

Finally, in the mid nineties we get acquainted with the work by Honig, Madhow and Verdù, and so we get convinced that interference mitigation could be really done and could work fine in a wireless satellite network. This was the beginning of the endeavor described in this book, where a small group of people from Academia, with the due technical support from a big semiconductor firm, where able to put together possibly the first ASIC-based CDMA interference mitigating detector ever. But this is just the beginning of a new era which I am sure will be as exciting as the previous decade.

Probably the most prominent Italian novelist, Alessandro Manzoni (1785–1873) used to modestly address his largely vast readership as *"my twenty-five readers"*. I am convinced that this book, too, will find (not the same!) twenty-five people that will enjoy and appreciate the spirit and lessons learnt during this remarkable adventure, as if they were themselves part of the team which carried out this exciting project financed by the ESA Technology Research Plan.

Riccardo De Gaudenzi

Head of the Communication Systems Section
European Space Research and Technology Centre
European Space Agency

Noordwijk (The Netherlands), July 2003

Chapter 1

INTRODUCING WIRELESS COMMUNICATIONS

"Life will not be the same after the wireless revolution". This is certainly true at the moment for countries in the Western world, and is going to be true in a few years for developing countries as well. So the aim of this Chapter is first to address the main terms of this revolution from the technical standpoint and to review the main second- and third-generation worldwide standards for wireless cellular communication, then to discuss how satellites can play a role in this scenario, and finally to show how this 'revolution' could have taken place through the tremendous technological progress of (micro-)electronics.

1. THE WIRELESS REVOLUTION

In many European countries the number of *wireless* access connections between the user terminals (cellular phones, laptops, palmtops, etc.) and the fixed, high capacity transport network has already exceeded the number of *wired* connections. Untethered communications and computing has ultimately become part of a lifestyle, and the trend will undoubtedly go further in the near future, with the commercialization of low cost *Wireless Local Area Networks* (WLANs) for the home. Round the corner we may also envisage pervasive, *ad hoc* wireless networks of sensors and user terminals communicating directly with each other via multiple hops, and without any need of support from the transport network.

The picture we have just depicted is what we may call the *wireless revolution* [Rap91]. Started in Europe in the early 90s, with the American countries lagging by a few years, it will probably come to its full evolution within the end of the first decade of the third century, to rise again in a second great tidal wave when the Asian developing countries will catch up [Sas98]. The real start of the revolution was the advent in Europe of the so called *2nd Gen-*

eration (2G), digital, pan-European cellular communication systems, the well known GSM (Global System for Mobile communications) [Pad95]. The explosive growth of cellular communications had already started with earlier analog systems, the so called *1ˢᵗ Generation* (1G), but the real breakthrough was marked by the initially slow, then exponential, diffusion of GSM terminals, fostered by continent wide compatibility through international roaming. In the United States the advent of 2G digital systems was somewhat slowed down by the co-existence of incompatible systems and by the consequent lack of a nation wide accepted unique standard. The two competing 2G American standards are the so called 'digital' AMPS (Advanced Mobile Phone System) IS-154 whose technology was developed with the specific aim of being compatible (as far as the assigned RF channels are concerned) with the pre-existing 1G analog AMPS system, and the highly innovative *Code Division Multiple Access* (CDMA) system IS-95.

In the second half of the 90s the GSM proved highly effective, boomed in Europe, and was adopted in many other countries across the whole world, including Australia, India, and most Asian countries. The initial European allocation of radio channels close to 900 MHz was paired by an additional allocation close to 1800 MHz (DCS-1800 system) that led to the tripling of system capacity. GSM techniques were also 'exported' to the United States under the label of PCS (Personal Communication Systems) with an allocation of channels close to 1900 MHz. At the turn of the century GSM, through its mature technology, started to be exploited as a true born digital system, delivering multimedia contents (paging, messaging, still images, and short videoclips). It is also being extended and augmented into a packet access radio network through the GPRS (Generalized Packet Radio Service) access mode (Figure 1-1), and will also be augmented to higher capacity through the EDGE (Enhanced Data rate for Global Evolution) technology. Both GPRS and EDGE are labeled '2.5G technologies', since they represent the bridge towards *3ʳᵈ Generation* (3G) systems which will be discussed later.

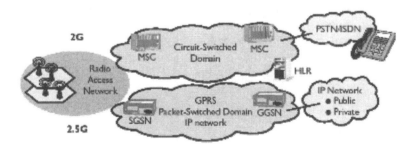

Figure 1-1. GSM/GPRS Network Architecture (http://www.gsm.org).

A similar evolution has taken place in the United States with CDMA IS-95 2G systems (Figure 1-2) [Koh95], [Gil91]. After a controversial start in the area of California, CDMA systems were early adopted in Brazil, Russia, and Korea. They soon evolved into an articulated family of different systems and technologies called cdmaOne, all based on the original standard IS-95 and its evolutions. After 'cellular CDMA' at 800 MHz was launched its PCS version at 1900 MHz was soon made available. Packet access was embedded into the system, and a standard for *fixed* radio terminals to provide fixed wireless access to the transport network was also added. We shall not insist further on the evolution of 1G and 2-2.5G systems in Japan, not to play again a well known song.

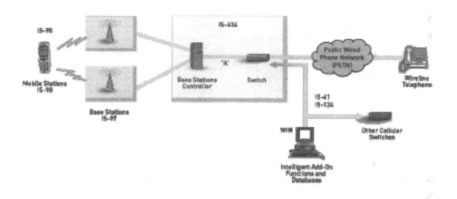

Figure 1-2. Network architecture of an IS-95 CDMA system (http://www.cdg.org).

At the dawn of the third millennium the ITU (International Telecommunications Union), based in Geneva, took the initiative of promoting the development of a universal 3G mobile/personal wireless communication system with high capacity and a high degree of inter-operability among the different network components, as depicted in Figure 1-3. Under the initiative IMT-2000 (International Mobile Telecommunications for the year 2000) [Chi92] a call for proposals was issued in 1997 to eventually set up the specifications and the technical recommendations for a universal system. At the end of the selection procedure, and in response to the different needs of the national industries, operators, and PTTs, two different non-compatible standards survived: UMTS (Universal Mobile Telecommunication System) for Europe and Japan, and cdma2000 for the USA. Both are based on a mixture of time and code division multiple access technologies. UMTS stems from a number of research projects carried out in the past by Europe and Japan (mainly FRAMES in Europe [Dah98] and CORE-A in Japan [Ada98]), whilst cdma2000, following a consolidated tradition in the standardization procedures in the United States, is a backwards compatible evolution of 2G CDMA [Kni98]. 3G systems are being developed at the time of the writing of this Chapter (early 2003), with Japan leading the group. Some are questioning the commercial validity of 3G systems (but this is something completely outside the scope of this book), other say that 3G will not reveal such a breakthrough as 2G systems have admittedly been. We will say more on 2G/3G systems in Section 1.2.

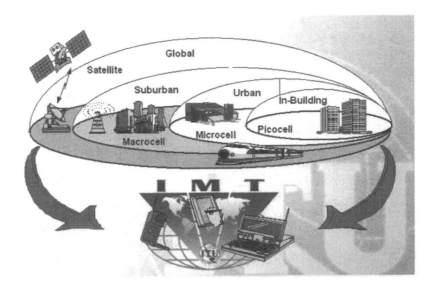

Figure 1-3. IMT-2000 system concept (http://www.itu.org).

The next wave of the wireless revolution may possibly come from WLANs [Nee99]. A WLAN is not just a replacement of a traditional wired LAN (such as the ubiquitous Ethernet in one of its 10/100/1000 Mbit/s versions). Many forecasts envisage, in fact, a co-existence between wired copper LANs to link fixed PCs within an office (or a building) and wireless networks (the WLANs) yielding high bit rate together with a certain support of mobility and handovers. With WLANs laptops, palmtops, possibly portable mp3 players and/or videoterminals, are all linked together, either via a central access point in a star topology (with immediate provision of connectivity with the fixed network), or directly with each other in an 'ad hoc', decentralized architecture. The former architecture is typical of IEEE 802.11a–b networks that are at the moment gaining more and more popularity; the latter is the paradigm of Bluetooth pico-nets/scatter-nets and of IEEE 802.15

'ad hoc' communications. WLANs are becoming the standard untethered connection for nomadic computing, and may challenge UMTS when full mobility is not a fundamental requirement. Still to come are WLANs for the home (such as HomeRF and similar products currently being developed) which belong more to the field of consumer electronics than to telecommunications. The physical layer of most efficient WLANs (802.11b) is based on multi-carrier modulation, borrowed from the fields of TV terrestrial broadcasting (Digital Video Broadcasting—Terrestrial, DVB-T) and Digital Subscriber Line access techniques (xDSL).

At the present time, people in the field of R&D for telecommunications speak of 4th *Generation* (4G) systems. At the moment nobody actually knows what 4G is going to be. The main trend for the physical interface is to combine CDMA for efficient access and frequency re-use, and multi-carrier transmission (as in WLANs) in order to cope best with radio propagation channel impairments, into a *Multi-Carrier CDMA* (MC-CDMA) signal transmission format. 4G networks are also expected to yield maximum capacity *and* flexibility. This means being able to integrate all of the scenarios mentioned above (traditional cellular systems for mobile communications, fixed wireless access systems, wireless LANs, 'ad hoc' network) into a single fully inter-operable, ubiquitous network.

The successful implementation of all of the different wireless systems' generations has relied, relies, and will rely on the formidable performance growth and cost/size decrease of *Very Large Scale Integrated* (VLSI) components. This fundamental enabling factor will be discussed in greater detail in Section 1.4, and is the pivot of all of the work described in this book.

2. 2G AND 3G WIRELESS COMMUNICATION SYSTEMS IN EUROPE AND THE USA

Both 2G and 3G systems are based on the concept of *cellular communications* and *channel frequency re-use* [Kuc91]. The concept of a cellular radio network is well known: the service area to be covered by a provider is split into a number of *cells* (usually hexagonal, as in Figure 1-4). Each cell is served by a Radio Base Station (RBS) which manages a number of channels whose center frequencies lie within the radio frequency spectrum allocated to that provider. In doing so, and with some specific techniques to be presented in a while, the same channels can be *re-used* in different cells, thus allowing them to serve a population of active users much larger than the mere number of allotted channels.

The main difference between GSM and IS-95 (the two European and American 2G digital cellular systems, respectively) lies in the way channel allocation within each cell is carried out. GSM is based on a mixture of *Time ivision Multiple Access* (TDMA) and *Frequency Division Multiple Access* (FDMA) to grant multiple access in the uplink, and on an equivalent TDM/FDM scheme for downlink multiplexing. Specifically, 8 TDMA channels are allocated to a single carrier, and the different carrier are spaced 200 kHz apart. Both with TDMA and with FDMA all channels can *not* be used in each cell, since that arrangement would give rise to excessive *inter-cell interference*. The latter comes from the possiblility of co-existent active channels on the same frequency and/or in the same time slot in adjacent cells. The solution of this issue is the technique of *channel frequency re-use*. As is shown in Figure 1-4, the cells on the coverage area are arranged into *clusters* (in the example, 7 cells/cluster). The total frequency band of the provider is then further split into chunks (as many chunks as cells in a cluster) of non-overlapping adjacent frequency channels (represented in Figure 1.4 by different shades of gray). The chunks are then permanently allocated to the cells of a cluster in such a way that cells using the same chunk of channels are sufficiently spaced so that the level of inter-cell interference is harmless to the quality of the radio link. This apparently places a limit on the overall network capacity in terms of channels/cell, which directly translates into a limitation of the served users/unit area.

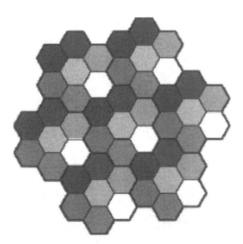

Figure 1-4. Cellular radio network with frequency re-use (http://www.cdg.org).

IS-95 is based on a different concept for multiplexing and multiple access. In the downlink the different channels, instead of different time slots or different carriers, are placed 'onto' different *spreading codes*, and they are transmitted at the same time and on the same carrier frequency. Such codes are taken from a set of *orthogonal functions* (the Walsh–Hadamard, or WH, sequences) in such a way that each channel can always be extracted from the 'mixture' of all of the downlink channels with no crosstalk from the others. This is the basic idea of *Code Division Multiplexing* (CDM) [DeG96] that will be discussed in greater detail in Chapter 2. The codes used for multiplexing are binary digital waveforms whose clock is faster than the data clock, and so they cause a *spectral spreading* of the data signal: CDM is inherently linked to *Spread Spectrum* (SS) modulation. The bandwidth increase factor is called the *spreading factor* of the spreading code, and is equal to 64 for IS-95. A similar concept is also used for the uplink, with a significant difference. The downlink WH sequences, also called *channelization* codes, stay orthogonal as long as they are *synchronous*. This is easily accomplished in the downlink, since the different tributary signals are physically co-located in the RBS. In contrast, the uplink signals coming from the different mobile user terminals cannot be easily synchronized, and thus the spreading codes of each user cannot be made synchronous with any of the others. Therefore, when accessing the radio channel the CDMA signals are asynchronous, and this causes a residual crosstalk on each channel coming from all the others. This crosstalk, called *Multiple Access Interference* (MAI), can be made sufficiently small by increasing the spreading factor of the uplink channels. In IS-95 the gross channel bandwidth is 1.25 MHz, whilst the maximum data rate is 19.2 kbit/s. Of course, the MAI creates either an impairment on the quality of the link, or a limitation in capacity. It is clear, in fact, that the MAI is proportional to the number of active users in a cell. If the number is too large the level of MAI is too high, and incoming calls may be dropped, causing a capacity boundary. This is something that is *not* experienced by FDMA/TDMA systems, wherein *intra-cell* MAI is totally absent owing to the orthogonality of uplink signals.

The real breakthrough of CDMA lies in the way channel re-use is handled. With CDMA, in fact, each cell, in addition to the *channelization* codes, is also given a unique *scrambling* code, so that each signal is 'doubly encoded'. The first level of coding (channelization) is necessary *within* the cell to make the different channels separable; the second level (scrambling) is necessary *between adjacent cells* to make signals arising from a different cell separable and not interfering; it is something similar to frequency re-use in FDMA/TDMA. But here there is actually *no* frequency re-use: thanks to the presence of the scrambling code, adjacent cells may use the same carrier frequency without creating an excessive level of *inter-cell* interference. Such an arrangement is represented in Figure 1-5, in which all cells are shaded the same way because all cells use the whole allocated bandwidth: it is the *universal frequency re-use*. The channelization code allows the re-use *all* of the channelization codes in all of the cells, thus increasing overall capacity in terms of active users per square km with respect to FDMA/TDMA with frequency re-use (Figure 1-4). Other features of CDMA (such as the use of channel codes to further increase capacity, the robustness to multi-path radio propagation, the possibility of performing seamless soft handover of communications between adjacent cells, and so on) have all contributed to make IS-95 and its evolutions a success.

Figure 1-5. CDMA universal frequency re-use (http://www.cdg.org).

But all 2G systems were substantially geared towards providing good-quality voice communications with limited data communication capabilities (just enough for paging and messaging services, and possibly for e-mail). Data channels were limited to a mere 9.6 or 14.4 kbit/s, which seemed satisfactory at the time of issuing the standards, but was revealed a few years later as patently inadequate for providing mobile Internet services and multimedia services in general. In the early 90s, therefore, different initiatives were taken in the United States, in Europe, and in Japan to develop an evolved 3G system with enhanced features: universal worldwide compatibility and roaming, increased capacity (up to 2 Mbit/s for fixed wireless services and 384 kbit/s with full mobility), support for co-existing multi-rate connections (typical of multimedia applications), fast packet access with 'always on' connections, and others. Such requirements led to the development of UMTS in Europe and Japan, and of cdma2000 in the Americas as mentioned in Section 1. They are both based on different forms of *wideband CDMA* (W-CDMA) technologies, where 'wide' is intended to refer to 2G system. The nominal bandwidth of a UMTS carrier is, in fact, 5 MHz, as opposed to the 1.25 MHz of IS-95. Both wideband CDMAs implement up-link/downlink full duplexing via *Frequency Division Duplexing* (FDD) and allocation of paired bands: for instance, European UMTS places the uplink in the 60 MHz band 1920–1980 MHz, and the downlink in the paired 2110–2170 MHz band. UMTS also encompasses an additional (optional) hybrid TDMA/CDMA mode with *Time Division Duplexing* (TDD) on the narrower unpaired bands 1900–1920 and 2010–2025 MHz, for a total of 35 MHz.

UMTS gained the headlines in newspapers worldwide around 2000 coinciding with the auctions of the frequency licenses, which took place in several European countries. During such auctions the cost of spectrum licenses in countries such as the United Kingdom and Germany reached unprecedented levels for mobile operators. The large expectation built around 3G mobile wireless systems, of which UMTS represents the European interpretation, has not materialized yet, unfortunately. However, despite the delays in the UMTS commercial roll out and the scepticism affecting the telecommunication world as a whole, UMTS will certainly play a key role in the development of multimedia wireless services. Although it is difficult to predict which kind of avenues 3G services will take, it is clear that the availability on the same device (i.e., the user terminal) of multimedia interactive services combined with accurate localization will open up astonishing possibilities for service providers. In addition to the current voice and short-message services, mobile users will be able to access the Internet at considerable peak speeds, download and upload documents, images, MP3 files, receive location dependent information, and so on. The mobile terminal functionalities will be greatly extended to make it a truly interactive Personal Digital Assistant (PDA) always connected to the Internet world and voice will just become one of the many multimedia services available at the user's fingertips.

3. THE ROLE OF SATELLITES IN 3G SYSTEMS

UMTS was originally intended to be made of a *terrestrial* and a *satellite* component (denoted as T-UMTS and S-UMTS, respectively) integrated in a seamless way. The economic troubles experienced by 2G satellite Global Mobile Personal Communication Systems (GMPCSs) such as Iridium and Globalstar have largely mitigated the investor enthusiasm for satellite based ventures. Those GMPCSs systems are based on large Low Earth Orbiting (LEO) 48-64 satellite constellations aimed to provide GSM-like services from hand held terminals with worldwide coverage. However, despite the severe financial difficulties encountered by LEO GMPCS, new regional systems based on geostationary (GEO) satellites, such as Thuraya and AceS, have been put in operation recently. Inmarsat, the first and most successful satellite mobile operator, intends to put into orbit new powerful GEO satellites providing UMTS-like services in the near future.

Since 1999 the European Space Agency (ESA) has carried out a number of system studies and technological developments to support the European and Canadian industry in the definition of S-UMTS component development strategy, identifying critical technological areas and promoting S-UMTS demonstrations [Bou02], [Cai99]. Particular effort has been devoted to the following key aspects:

i)   study and optimization of system architectures providing appealing features to possible operators;

ii)   design, testing, and standardization within international bodies of an air interface with maximum commonality with terrestrial UMTS;

iii)   development and validation of real time demonstrators for laboratory and over the air S-UMTS experiments;

iv)   design and development of large reflector antennas;

v)   design and prototyping of an advanced high throughput On Board Processor (OBP) for future mobile missions;

vi)   networking studies and simulations.

Furthermore, ESA, in strict cooperation with the European Commission, has promoted the creation of the Advanced Mobile Satellite Task Force (ASMS-TF) which has now attracted more than 45 worldwide key players working together to define and defend the role of satellite in 3G mobile and beyond.

The main new opportunities identified for S-UMTS and more in general for Advanced Mobile Systems by ESA studies and the ASMS-TF are:

Development of Direct Mobile Multicasting & Broadcasting: addressing mainly the consumer market, as well as specific corporate markets. This kind of system would overlay the terrestrial cellular networks by implementing 'point to multi point' services in a more efficient manner, and would complement terrestrial digital broadcast networks by extending and completing their geographical coverage.

Extension of the wireless mobile terrestrial networks: i.e., coverage extension, coverage completion, global roaming, and rapid deployment. It is expected that in this market segment satellite mobile systems will require a much higher degree of integration with terrestrial infrastructures than for extensions of existing mobile satellite systems.

A possible S-UMTS system architecture encompassing both opportunities listed above is presented in Figure 1-6. In this case the satellite provides 'Point to Point' (P2P) unicast services for T-UMTS geographical complement, as well as T-UMTS complementary broadcast and multicast services.

The latter will profit from the presence of a terrestrial gap filler network which will boost the Quality of Service (QoS) in densely populated areas where the satellite signal is often too weak to be received. T-UMTS coverage extension may reveal an attractive solution for providing UMTS services to regions with low density population. The satellite user link is at S band (2 GHz) as for T-UMTS, whilst the service link between the gateway stations and the satellite is at Ka band (20–30 GHz). The service links are mapped to the user beams by means of a bent pipe satellite transponder. For broadcast services, only the forward link (i.e., the gateway → satellite → user link) is implemented. For geographical T-UMTS extension a reverse link will be also implemented to provide a user → satellite → gateway connection. Terrestrial gap fillers receive the broadcast satellite signal directly from the satellite on a dedicated Ka band link and perform frequency downconversion on the same S-UMTS 2 GHz bands as above.

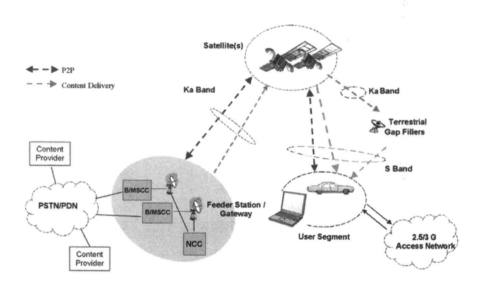

Figure 1-6. Satellite-UMTS Architecture.

Terrestrial UMTS coverage extension calls for the generation of a large number of relatively small radio beams (compared to usual satellite beams, see Figure 1-7 as an example) generated by very large antenna reflectors (20–25 m). This allows the achievement of a considerable frequency re-use factor, similarly to what is done in a terrestrial cellular network (a cell is replaced here by a beam). But, as stated above, S-UMTS may also complement terrestrial UMTS in terms of services. This is why the concept of direct mobile digital multimedia multicasting and broadcasting is considered to be the best chance for satellite systems to access the mass mobile market. Terrestrial UMTS networks are based on very small cells (of a few km radius) which allow the provision of high peak rates (up to a few hundreds kbit/s) and attain a large frequency re-use. This is possible thanks to the CDMA technology which is at the core of the UMTS radio interface. However, the technical solution becomes very inefficient when the UMTS networks have to be used to transmit the same information to many users (e.g., video clips containing highlights of sport events, financial information, broadcasting of most accessed web pages, etc.). Broadcast information may be locally stored in the user terminal (*cacheing*) and accessed when required by the user. In this case a satellite broadcast layer with large cell sizes (Figure 1-8) on top of T-UMTS will provide a cheaper solution for this kind of services. Satellite systems have a major advantage in broadcasting information since a single satellite can cover regions as large as Europe or USA. Good quality of service can be achieved by means of powerful error correction techniques and by an integrated terrestrial gap filler network.

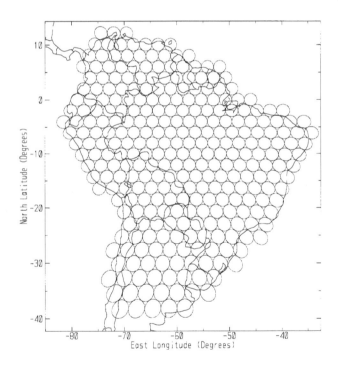

Figure 1-7. Sample multi-beam satellite footprint (courtesy of Alenia Spazio, Italy).

In all cases the efficient use of the very scarce and expensive spectrum resources causes the S-UMTS system to operate under heavy interference conditions. The interference is mainly caused by other satellite beams which are re-using the same frequency (see Figure 1-9). In the case of terrestrial gap fillers other co-frequency transmitters also generate further co-channel interference. So the issue of co-channel interference mitigation to increase system capacity is pivotal to the economical development and deployment of (S-)UMTS.

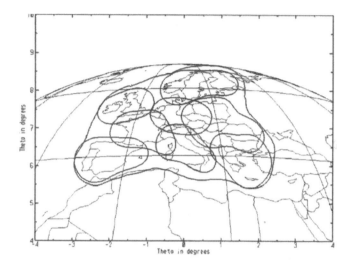

Figure 1-8. Sample broadcast footprint (courtesy of Alenia Spazio, Italy).

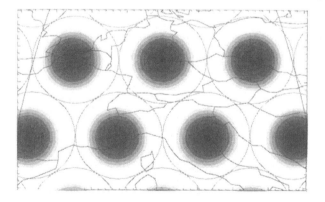

Figure 1-9. Interference pattern for a European multi-beam UMTS satellite (courtesy of Alenia Spazio, Italy).

4.   VLSI TECHNOLOGIES FOR WIRELESS COM-
MUNICATION TERMINALS

It is an everyday experience of life to buy the best and most expensive cellular phone at one's retail store and after one year or so to find the same item at half that price. This is just the result of the celebrated Moore's law: the number of transistor on a chip with a fixed area roughly doubles every year and a half. Hence the price of microelectronics components halves in the same period, or, the power of VLSI (Very Large-Scale Integrated) circuits doubles over the same 18 months.

As a matter of fact, over the last few decades Moore's prediction has been remarkably prescient. The minimum sizes of the features of CMOS (Complementary Metal Oxide–Semiconductor) transistors have decreased on average by 13% per year from 3 μm in 1980 to 0.13 μm in 2002, die areas have increased by 13% per year, and design complexity (measured by the number of transistors on a chip) has increased at an annual growth rate of 50% for Dynamic Random Access Memories (DRAMs) and 35% for microprocessors. Performance enhancements have been equally impressive. For example, clock frequencies for leading edge microprocessors have increased by more than 30% per year. An example related to transistor count of Intel microprocessors is reported in Figure 1-10.

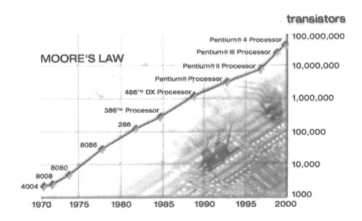

Figure 1-10. Moore's law and Intel microprocessors (courtesy of Intel).

This enormous progress in semiconductor technology is fueling the growth in commercial wireless communications systems. New technologies are being spurred on by the desire to produce high performance, low power, small size, low cost, and high efficiency wireless terminals. The complexity of wireless communication systems is significantly increasing with the application of more sophisticated multiple-access, digital modulation and processing techniques in order to accommodate the tremendous growth in the number of subscribers, thus offering vastly increased functionality with better quality of service. In Figure 1-11 we have illustrated the processor performance (according to Moore's Law) together with a qualitative indication of the algorithm complexity increase (approaching the theoretic performance limits imposed by Shannon's theory), which leaps forward whenever a new wireless generation is introduced, as well as the available battery capacity, which unfortunately increases only marginally [Rab00].

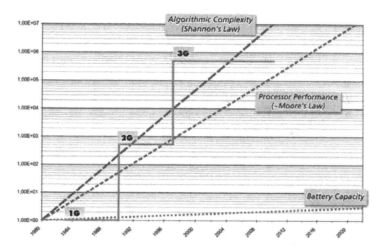

Figure 1-11. Moore's law, system complexity and battery capacity [Rab00].

It appears that system complexity grows faster than Moore's Law, and so a 'brute force' use of the available processing power (GIPS/s) in a fully programmable implementation is not sufficient; often dedicated hardware accelerators are required. Furthermore, taking the battery capacity limit into account, the use of dedicated hardware becomes mandatory to reduce power consumption. This is the fundamental trade off between energy efficiency (i.e., the number of operations that can be performed for a given amount of energy) and flexibility (i.e., the possibility to re-use a single design for multiple applications) which is clearly illustrated in Figure 1-12 for various implementation styles [Rab00]. An amazing three orders of magnitude vriation of energy efficiency (as measured in MOPS/mW) can be observed between an ASIC (Application Specific Integrated Circuit) style solution and a fully programmable implementation on an embedded processor. The differences are mostly owed to the overhead that comes with flexibility. Application specific processors and configurable solutions improve energy efficiency at the expense of flexibility. The most obvious way of combining flexibility and cost efficiency is to take the best from different worlds: computationally intensive signal processing tasks are better implemented on DSP (Digital Signal Processor) cores or media processor cores than on a microprocessor core, whilst the opposite is true for control tasks.

As shown in Figure 1-13, a typical wireless transceiver combines a data pipe, which gradually transforms the bit serial data stream coming from the Analog to Digital Converter (ADC) into a set of complex data messages, and a protocol stack, that controls the operation of the data pipe. Data pipe and protocol stack differ in the kind of computation that is to be performed, and in the communication mechanisms between the functional modules. In addition, the different modules of the data and control stacks operate on time and data granularities which vary over a wide range. The conclusion is that a heterogeneous architecture which optimally explores the 'flexibility-power-performance-cost' design space is the only viable solution of handling the exponentially increasing algorithmic complexity (which is mainly owed to multiple standards, adaptability and increased functionality) and the battery power constraint in wireless terminals. Figure 1-14 shows a typical heterogeneous System on a Chip (SoC) architecture employing several standard as well as application specific programmable processors, on chip memories, bus based architecture, dedicated hardware co-processor, peripherals, and Inpu/Output (I/O) channels.

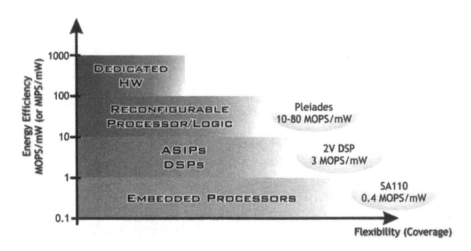

Figure 1-12. Trading off flexibility versus energy efficiency (in MOPS/mw or million of operation per mJ of energy) for different implementation styles [Rab00].

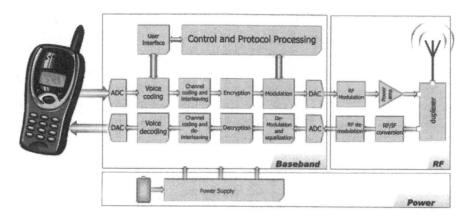

Figure 1-13. Functional components of a wireless transceiver.

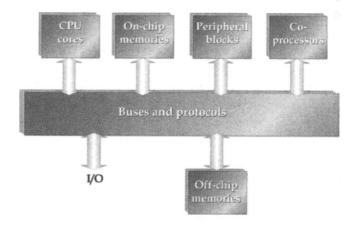

Figure 1-14. Typical heterogeneous System-on-Chip platform.

Now that the main architecture of the terminal is decided, the subsequent key problem is how to map the system/algorithm onto the various building blocks of a heterogeneous, configurable SoC architecture (hardware and software) within given constraints of cost and time to market.

An extensive profiling/analysis of the application/algorithm in the early algorithmic design phases can help to determine the required bounds on performance and flexibility, or to outline the dominant computational pattern and explore data transfer and storage communications. This step is both tedious and error prone if carried out fully manually, and so new design methodologies have to be provided to bridge the gap between algorithmic development and cost effective realization. There is a need for fast guidance and early feedback at the algorithm level, without going all the way down to assembly code or hardware layout (thus getting rid of long design cycles). Only when the design space has been sufficiently explored at a high level, and when a limited number of promising candidates have been identified, a more thorough and accurate evaluation is required for the final hardware/software partitioning. Most importantly, the optimum system is always the result of a joint, truly, interactive architecture–algorithm design. A better algorithm (even the best) from a communication performance standpoint may not correspond to a suitable computational/communication architecture. Since no single designer can adequately handle algorithms, design methodologies and architectures, a close interaction between designers (the system/communication engineer and the VLSI/chip architect) and design teams is required to master such a complex SoC design space.

Therefore, the designer's efficiency must be improved by a new design methodology which benefits from the re-use of Intellectual Property (IP) and which is supported by appropriate tools that allow the joint design and verification of heterogeneous hardware and software. Particularly, owing to the exponential increase of both design gate counts and verification vectors, the verification gap grows faster than the design size by a factor of 2/3 according to the International Technology Roadmap for Semiconductor (ITRS) road map.

This is the well known design productivity challenge that has existed for a long time. Figure 1-15 shows how Integrated Circuits (ICs) complexity (in logic transistors) is growing faster than the productivity of a design engineer, creating a 'design gap'. One way of addressing this gap is to steadily increase the size of the design teams working on a single project. We observe this trend in the high performance processor world, where teams of more than a few hundred people are no longer a surprise. This approach cannot be sustained in the long term, but fortunately, about once in a decade we witness the introduction of a novel design methodology that creates a step function in design productivity, helping to bridge the gap temporarily. Looking back over the past four decades, we can identify a certain number of productivity leaps. Pure custom design was the norm in the early integrated circuits of the 1970s. Since then programmable logic arrays, standard cell, macrocells, module compilers, gate arrays, and reconfigurable hardware have steadily helped to ease the time and cost of mapping a function onto silicon. Today semiconductor technology allows the integration of a wide range of complex functions on a single die, the SoC concept already mentioned. This approach introduces some major challenges which have to be addressed for the technology to become a viable undertaking: *i*) very high cost of production facilities and mask making (in 0.13 μm chips, mask costs of $600,000 are not uncommon); *ii*) increase performance predictability reducing the risk involved in complex SoC design and manufacturing as a result of deep sub-micron (0.13 μm and below) second order effects (such as crosstalk, electro-migration and wire delays which can be more important than gate delays). These observations, combined with an intense pressure to reduce the time to market, requires a design paradigm shift comparable with the advent of the driving of ASIC design by cell libraries in the early 1980s, to move to the next design productivity level by further raising the level of abstraction. To this aim, recently the use of platforms at all of the key articulation points in the SoC design has been advocated [Fer99]. Each platform represents a layer in the design flow for which the underlying subsequent design flow steps are abstracted. By carefully defining the platforms' layers and developing new representations and associated transitions from one platform to the next, an economically feasible SoC design flow can be realized. Platform based design provides a rigorous foundation for design re-use, 'correct by construction' assembly of pre-designed and pre-characterized components (versus full custom design methods), design flexibility (through an extended use of reconfigurable and programmable modules) and efficient compilation from specification to implementations. At the same time it allows us to trade off various components of manufacturing, Non-Recurrent Engineering (NRE) and design costs while sacrificing as little potential design performance as possible. A number of companies have already embraced the platform concept in the design of integrated embedded systems. Examples are the Nexperia platform by Philips Semiconductor [Cla00], [Gra02], the Gold platform by Infineon [Hau01], and the Ericsson Mobile Platform by Ericsson [Mat02].

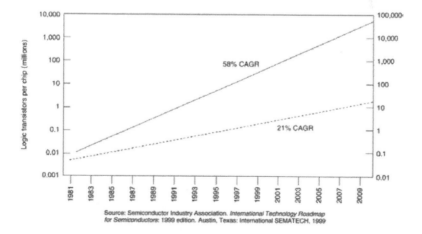

Source: Semiconductor Industry Association. *International Technology Roadmap for Semiconductors*: 1999 edition. Austin, Texas: International SEMATECH, 1999

Figure 1-15. The design productivity gap, showing the different Compound Annual Growth Rates (CAGRs) of technology (in logic transistors per chip) and design productivity (in transistors designed by a single design engineer per month) over the past two decades.

Chapter 2

BASICS OF CDMA FOR WIRELESS COMMUNICATIONS

Is the reader familiar with the basic concepts in CDMA communications? Then he/she can safely skip the three initial Sections of this Chapter. If he/she is not, he/she will find there the main issues in generation and detection of a CDMA signal, and the basic architecture of a DSP-based CDMA receiver. But even the more experienced reader will benefit from the subsequent three sections of this Chapter, which deal with the use of CDMA in a satellite mobile network (with typical numerical values of the main system parameters), with the relevant techniques for interference mitigation (cancellation), and with the specifications of the case considered in the book and referred to as MUSIC (Multi USer and Interference Cancellation).

1. NARROWBAND AND WIDEBAND DIGITAL MODULATIONS

The generic expression of a linear band pass modulated signal $s(t)$ is

$$s(t) = s_I(t) \cdot \cos(2\pi f_0 t) - s_Q(t) \cdot \sin(2\pi f_0 t), \tag{2.1}$$

where f_0 is the carrier frequency, whilst $s_I(t)$ and $s_Q(t)$ are two baseband signals which represent the *In phase* (I) and the *Quadrature* (Q) components of the modulated signal, respectively. A more compact representation of the I/Q modulated signal (2.1), provided that the carrier frequency f_0 is known, is its *complex envelope* (or *baseband equivalent*) defined as follows

$$\tilde{s}(t) = s_I(t) + js_Q(t).$$

(2.2)

The relation between the complex-valued baseband equivalent and the real-valued band pass modulated signal is straightforward

$$s(t) = \Re\{\tilde{s}(t) \cdot e^{j2\pi f_0 t}\}.$$

(2.3)

The normalized power of the baseband components $s_I(t)$ and $s_Q(t)$ are

$$P_{s_I} = \mathrm{E}\{s_I^2(t)\}, \ P_{s_Q} = \mathrm{E}\{s_Q^2(t)\},$$

(2.4)

so that from (2.2) the power of the complex envelope $\tilde{s}(t)$ is

$$P_{\tilde{s}} = \mathrm{E}\{|\tilde{s}(t)|^2\} = P_{s_I} + P_{s_Q}.$$

(2.5)

From (2.1) we also find that the Radio Frequency (RF) power of the modulated signal $s(t)$ is given by

$$P_s = \mathrm{E}\{s^2(t)\} = \frac{P_{s_I} + P_{s_Q}}{2} = \frac{P_{\tilde{s}}}{2}.$$

(2.6)

Assume now that the information data source is generating a stream of information bearing binary symbols (*bits*) $\{u_m\}$ running at a rate $R_b = 1/T_b$, where T_b is the bit interval. The information bits are mapped onto a *constellation* of W symbols (represented as a set of points in the complex plane) and each complex symbol is then labeled by a 'word' of $\log_2(W)$ bits. This mapping generates a stream of complex-valued symbols $\{\tilde{d}_k\}$ with

$$\tilde{d}_k = d_{I,k} + jd_{Q,k}$$

(2.7)

running at the rate $R_s = 1/T_s$, where T_s is the symbol, or signaling, interval, $R_s = R_b/\log_2(W)$. The symbols are subsequently shaped by a filter with a T_s energy impulse response $g_T(t)$ so as to obtain the following (baseband equivalent) data modulated signal

$$\tilde{s}(t) = \sqrt{\frac{2P_s}{A_d^2}} \cdot \sum_{k=-\infty}^{\infty} \tilde{d}_k \cdot g_T(t - kT_s).$$

(2.8)

The equation above is the general expression of a linear digital modulation where P_s is the RF power of the modulated signal (2.1), as defined in (2.6), and $A_{\tilde{d}}^2$ is the mean squared value of the data symbols

$$A_{\tilde{d}}^2 = \mathrm{E}\left\{\left|\tilde{d}_k\right|^2\right\} = A_{d_I}^2 + A_{d_Q}^2 \qquad (2.9)$$

with

$$A_{d_I}^2 = \mathrm{E}\left\{d_{I,k}^2\right\}, \quad A_{d_Q}^2 = \mathrm{E}\left\{d_{Q,k}^2\right\}. \qquad (2.10)$$

The block diagram of the linear I/Q modulator described in (2.8), is shown in Figure 2-1, where we have introduced the amplitude coefficient

$$A = \sqrt{2P_s / A_{\tilde{d}}^2} \qquad (2.11)$$

and its relevant baseband complex equivalent, as in (2.2), is shown in Figure 2-2. The linear *I/Q demodulator* to recover the digital data from signal (2.3) is shown in Figure 2-3, where the (ideal) low pass filters $H(f)$ detect the baseband components $s_I(t)$ and $s_Q(t)$, and suppress double frequency components (*image spectra*) arising from the previous mixing process. In all of the figures in this chapter and in the remaining part of the book, thick lines denote complex-valued signals.

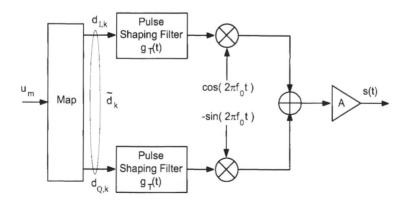

Figure 2-1. Block diagram of a linear I/Q modulator.

Figure 2-2. Baseband block diagram of a linear I/Q modulator.

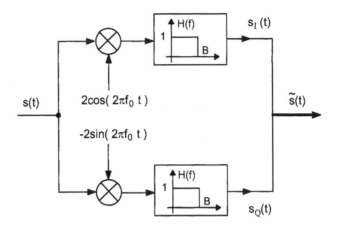

Figure 2-3. Block diagram of a linear I/Q demodulator (B is the signal bandwidth).

Probably the most popular truly I/Q modulation format is *Quadrature Amplitude Modulation* (QAM) with a square W-point constellation (W-QAM). Assuming that $W = w^2$, we have

$$d_{I,k}, d_{Q,k} \in \left\{ \pm 1, \pm 3, \ldots, \left(\sqrt{W} - 1\right) \right\} = \left\{ \pm 1, \pm 3, \ldots, \pm (w - 1) \right\}, \qquad (2.12)$$

and therefore

$$A_{\tilde{d}}^2 = 2 \cdot \frac{W - 1}{3}. \qquad (2.13)$$

Whilst W-QAM is widely used in wireline modems, satellite communications more often rely on *Phase Shift Keying* (PSK) constellations. A W-PSK constellation is defined by

$$\tilde{d}_k \in \left\{ e^{jm\frac{2\pi}{W}} \right\}, \qquad (2.14)$$

so that

$$A_{\tilde{d}}^2 = 1. \qquad (2.15)$$

The two classes of PSK and QAM signals have a common element, since the two formats of 4-QAM and 4-PSK (or *Quadrature Phase Shift keying, QPSK*) are equivalent.

It can be shown [Pro95] that the Power Spectral Density (PSD) of a base-band digital signal (expressed in $[V^2 / Hz]$) is

$$\mathcal{P}_{\tilde{s}}(f) = \frac{1}{T_s} \cdot \mathcal{P}_{\tilde{d}\tilde{d}}(f) \cdot \left| G_T(f) \right|^2, \qquad (2.16)$$

where $G_T(f)$ is the frequency response of the pulse shaping filter (whose impulse response is $g_T(t)$), and where we have introduced the so called *data spectrum*

$$\mathcal{P}_{\tilde{d}\tilde{d}}(f) = \sum_{k=-\infty}^{\infty} R_{\tilde{d}\tilde{d}}(k) \cdot e^{-j2\pi k T_s f} \qquad (2.17)$$

as the Fourier transform of the (discrete time) data autocorrelation sequence

$$R_{\tilde{d}\tilde{d}}(k) = \mathrm{E}\left\{ \tilde{d}_m \cdot \tilde{d}_{m+k}^* \right\}. \qquad (2.18)$$

Assuming that the symbol rate R_s is much smaller than the carrier frequency f_0, the PSD of a modulated signal such as (2.1) can be obtained from that of the relevant complex envelope, as follows

$$\mathcal{P}_s(f) = \frac{1}{4} \cdot \left[\mathcal{P}_{\tilde{s}}(f - f_0) + \mathcal{P}_{\tilde{s}}(f + f_0) \right]. \qquad (2.19)$$

The power of the baseband signal is then

$$P_{\tilde{s}} = \int_{-\infty}^{\infty} \mathcal{P}_{\tilde{s}}(f) \, df, \qquad (2.20)$$

whilst for the band pass signal we obtain from (2.19)

$$P_s = \int_{-\infty}^{\infty} P_s(f)\, df = \frac{P_s}{2} \qquad (2.21)$$

in agreement with (2.6).

The most naive data pulse is the rectangular shaping, whereby we have

$$g_T(t) = \mathrm{rect}\left(\frac{t}{T_s}\right) \qquad (2.22)$$

and

$$G_T(f) = T_s \cdot \mathrm{sinc}(fT_s) \qquad (2.23)$$

From the expression of the baseband PSD (2.16) we have that in the case of rectangular shaping the signal bandwidth measured at the first spectral null is given by

$$B_{1st\ null} = R_s = 1/T_s . \qquad (2.24)$$

When bandwidth comes at a premium (as is always the case in wireless communications) some form of band limiting is in order, and so rectangular pulses are no longer used. Let us therefore introduce the *Nyquist's Raised Cosine* (RC) pulse defined as follows

$$g_{Nyq}(t) = \frac{\cos(\pi \alpha t / T_s)}{1 - (2\alpha t / T_s)^2} \cdot \mathrm{sinc}\left(\frac{t}{T_s}\right) \qquad (2.25)$$

whose spectrum is (see Figure 2-4)

$$G_{Nyq}(f) = \begin{cases} T_s & -\left(\dfrac{1-\alpha}{2T_s}\right) \le f \le \dfrac{1-\alpha}{2T_s}, \\[4mm] \dfrac{1}{2}\left\{1 + \cos\left[\dfrac{\pi\left(|f| - \dfrac{1-\alpha}{2T_s}\right)}{\dfrac{\alpha}{T_s}}\right]\right\} & \dfrac{1-\alpha}{2T_s} \le |f| \le \dfrac{1+\alpha}{2T_s}, \\[4mm] 0 & |f| \ge \dfrac{1+\alpha}{2T_s}. \end{cases} \qquad (2.26)$$

where α (with $0 \le \alpha \le 1$) is the *roll off factor*. The most popular band limited transmission pulse shape is the (T_s energy) *Nyquist's Square Root Raised Cosine* (SRRC), given in the frequency domain by

$$G_T(f) = \sqrt{T_s \cdot G_{Nyq}(f)}. \tag{2.27}$$

The bandwidth occupancy of a baseband digital signal with SRRC shaping is therefore (see Figure 2-4)

$$B_{max} = (1+\alpha) \cdot \frac{R_s}{2} = \frac{1+\alpha}{2T_s} \tag{2.28}$$

and its −3 dB bandwidth is

$$B_{-3dB} = \frac{R_s}{2} = \frac{1}{2T_s}. \tag{2.29}$$

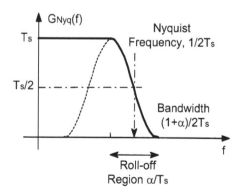

Figure 2-4. Spectrum of the Nyquist's SRRC pulse.

In the case of *Spread Spectrum* (SS) systems the bandwidth occupancy of the transmitted signal is intentionally augmented well beyond the value required for conventional narrowband transmissions, and this can be essentially accomplished in two ways.

The first one, at least from an historical perspective, dates back to 1941 and was conceived for providing secure military communications during WWII. The basic idea underlying this early spread spectrum concept consists in partitioning the overall spectrum into K separate channels or bands (also referred to as *frequency bins*), all having the same bandwidth as the signal to be transmitted. Spectrum spreading is achieved by transmitting the (conventional) narrowband signal into one bin for a given time interval T_{hop} , and then by randomly changing the carrier frequency of the modulator so as to place the signal spectrum in another bin, and so forth. The final result is that the signal spectrum 'hops' from bin to bin with an apparently casual pattern to escape hostile jamming and/or eavesdropping by unauthorized listeners. Owing to such a feature, this technique has been named *Frequency Hopping Spread Spectrum* (FH/SS). Actually, the pattern of the frequency bins is not truly casual, but it is instead *pseudo-random* (i.e., apparently random) and repeats every $K \cdot T_{hop}$ seconds. The (periodic) frequency pattern is made known also to the authorized receiver which is then capable of tracking the transmitted carrier.

It turns out that the total bandwidth occupancy of this SS system is K times that of the original narrowband signal. It is also apparent that the robustness of a FH/SS system against interference and unauthorized detection increases with the length K of the frequency hopping pattern, i.e., with the total occupied bandwidth. However, owing to strict requirements concerning oscillator frequency stability and switching rate, FH/SS does not find significant applications in multiple access commercial systems for mobile and cellular communications, and therefore it will not be considered any further in the following. The interested reader can find more details about the origins of SS in [Sch82] and about FH/SS transmssion in [Sim85], [Dix94].

An alternative way of generating a SS signal consists of the direct multiplication, in the time domain, of the information bearing symbols, running at the rate $R_s = 1/T_s$, with a sequence of binary symbols (*chips*) running at a much higher rate $R_c = 1/T_c$ (with T_c the chip interval), and with a repetition period L [Dix94], [Pic82], [Sim85]. This multiplication, which is carried out before transmit pulse shaping takes place, produces a stream of high rate symbols running at the rate R_c – the resulting signal to be transmitted turns out to have a bandwidth occupancy wider than that of conventional modulation schemes. For this reason the sequence of high rate chips is also referred to as *spreading sequence* or *spreading code*, and this wideband transmission technique is called *Direct Sequence Spread Spectrum* (DS/SS). This is the most used spread spectrum signaling technique used in commercial communication systems. In the following we will therefore restrict our attention to DS/SS transmission schemes.

Let us focus now on the analytical description of a DS/SS signal format and on the relevant features [DeG96]. In the simplest arrangement the spreading sequence has real-valued chips c_k selected at random in the binary alphabet $\{-1,1\}$ (this is what is done, for instance, in the uplink of the American cellular CDMA systems IS-95 or cdmaOne). Thus the spreading code is random non-periodic. It may also be expedient to have a *periodic* pseudo-random spreading code with a repetition length (repetition period) equal to L chips. When this *short code* arrangement is adopted (as for instance in the UMTS downlink), the repetition period of the spreading sequence usually coincides with one symbol interval, i.e., $T_s = L \cdot T_c$. The resulting DS/SS signal for the short code format is

$$\tilde{s}^{(SS)}(t) = A \cdot \sum_{k=-\infty}^{\infty} \tilde{d}_{\{k\}_L} \cdot c_{|k|_L} \cdot g_T(t - kT_c),$$ (2.30)

where we have introduced the following operators

$$\{k\}_L = \text{int}\left(\frac{k}{L}\right), \quad |k|_L = k \bmod L.$$ (2.31)

Figure 2-5 depicts the DS/SS transmitter described by (2.30), in which the amplitude coefficient A has the same meaning as in Figures 2-1 and 2-2. It can be shown, in fact, that the PSD of the DS/SS signal is given by the convolution between the narrowband PSD of the data and the wideband PSD of the spreading sequence, and that the power of the DS/SS signal is exactly the same as the power P_s of the narrowband signal.

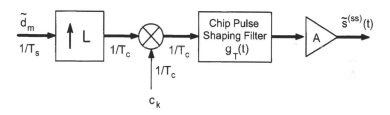

Figure 2-5. Baseband block diagram of a DS/SS transmitter.

It is apparent that the sections of the DS/SS transmitter in Figure 2-5 are the same as those in Figure 2-2, provided that the *chip* rate $R_c = 1/T_c$ is exchanged with the symbol rate R_s. In particular, the *chip* shaping filter in Figure 2-5 outputs a new T_c energy pulse $g_T(t)$ every T_c seconds. The index k ticks at the rate R_c, the chips' subscript $|k|_L$ repeats every L chip intervals, and the data symbols subscript $\{k\}_L$ ticks at the rate $LR_c = R_s$ as it should. In the case of a DS/SS transmission with rectangular shaping, the pulse (2.22) then becomes

$$g_T(t) = \text{rect}\left(\frac{t}{T_c}\right),$$ (2.32)

and the bandwidth occupancy (2.24) must be modified accordingly,

$$B_{1st\ null}^{(SS)} = R_c = \frac{1}{T_c}.$$

(2.33)

Apart from some particular applications, mainly military and the Global Positioning System (GPS), the spectrum of the DS/SS signal will be strictly limited. This can be achieved by resorting to Nyquist's SRRC shaping of the chip pulses. These pulses are described by (2.25)–(2.27) where the symbol duration T_s must be replaced with the chip interval T_c. Similarly the bandwidths (2.28) and (2.29) modify as follows

$$B_{max}^{(SS)} = (1+\alpha) \cdot \frac{R_c}{2} = \frac{1+\alpha}{2T_c},$$

(2.34)

$$B_{-3dB}^{(SS)} = \frac{R_c}{2} = \frac{1}{2T_c},$$

(2.35)

respectively. After spreading and chip pulse shaping, the baseband signal (2.30) is eventually upconverted to RF and transmitted using a pair of I/Q carriers as in Figure 2-1.

2.   PROPERTIES OF SPREAD SPECTRUM SIGNALS

We generalize now the DS/SS signal format (2.30), by allowing in particular that the code repetition length L be different from the so called *spreading factor*

$$M = \frac{T_s}{T_c} = \frac{R_c}{R_s}$$

(2.36)

We obtain thus the following expression [DeG96]

$$\tilde{s}^{(SS)}(t) = A \cdot \sum_{k=-\infty}^{\infty} \tilde{d}_{\{k\}_M} \cdot c_{|k|_L} \cdot g_T(t - kT_c),$$

(2.37)

where, according to the definitions (2.31), the subscript $\{k\}_M$ of the data symbols is now updated every M ticks of the index k. This is the most general form of a DS/SS signal.

We also remark that for any of the bandwidth definitions presented above we have

$$M = \frac{B^{(SS)}}{B}. \tag{2.38}$$

Therefore the parameter M can be seen as the ratio between the bandwidth occupancy of the SS signal $B^{(SS)}$ and the bandwidth B of the conventional narrowband signal. This is why we called M the *spreading factor*, that represents one of the main parameters characterizing the properties of a SS signal. Figure 2-6 shows the PSD of a code for DS/SS transmissions with spreading factor $M = 8$ compared with the PSD of a conventional narrowband signal with rectangular pulse shaping. If one symbol interval spans exactly an integer number n of the spreading sequence repetition periods,

$$T_s = n \cdot L \cdot T_c, \tag{2.39}$$

then the spreading sequence is termed *short code* and the spreading factor is $M = n \cdot L$. Otherwise, if the repetition period of the spreading sequence is much longer than the symbol duration, $L \cdot T_c \gg T_s$, then the sequence is called *long code* and the spreading factor is $M \ll L$. As a particular case of short code spreading we have the simplest configuration with $n = 1$ mentioned in the previous section, where the code repetition period is exactly coincident with one symbol interval, i.e., $M = L$.

One other important parameter of SS signals is the *processing gain* G_p, which is often confused with the spreading factor, and which actually has to do with the anti-jam capability of the receiver. The processing gain is defined as the ratio between the chip rate and the information bit rate

$$G_p = \frac{R_c}{R_b} = \frac{T_b}{T_c}. \tag{2.40}$$

Recalling that $R_s = R_b / \log_2(W)$, it is easy to derive the relationship between spreading factor and processing gain

$$M = G_p \cdot \log_2(W). \tag{2.41}$$

A widely used method for the generation of a binary spreading sequence $\{c_k\}$ is based on an m stage *Linear Feedback Shift Register* (LFSR) used as a binary sequence generator. The feedback taps of the circuit are properly set to yield a sequence with particular features which are desirable for SS transmissions [Dix94], [Pic82], [Sim85], [Din98]. According to the polynomial theory [Pet72], it is possible to select the feedback taps so as to obtain a sequence with the longest possible repetition period $L = 2^m - 1$ and good correlation properties [Sar80], the so called *maximal length* or *m-sequences*. Tables for the configurations of m-sequence generators are easily found in the literature [Dix94], [Pic82]. In an m-sequence the number of logical '0's and '1's is $2^{m-1} - 1$ and 2^{m-1}, respectively [Dix94]. Since in all practical cases we have $L \gg 1$ the occurrence rate of each symbol is virtually the same within each period, so that such sequences are called *balanced*. A deeper analysis of the location of '0's and '1's within one period also reveals that the blocks of consecutive identical symbols (the so called *runs*) follow a given distribution that makes the appearance of the sequence similar to that of a truly random one [Dix94]. Owing to these properties a sequence of this kind is also named *Pseudo-Random Binary Sequence* (PRBS).

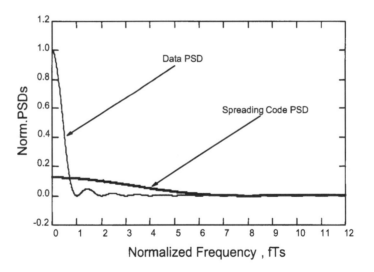

Figure 2-6. PSD of a spreading code for DS/SS transmissions, M=8.

Concerning correlation properties, we define first the following *periodic cross-correlation* sequence between the two generic binary sequences $\{a_k\}$ and $\{b_k\}$, $a_k, b_k \in \{-1, +1\}$

$$R_{ab}(i) = \frac{1}{L} \cdot \sum_{k=0}^{L-1} a_k \cdot b_{|k+i|_L} .$$

(2.42)

We will also denote the zero lag correlation as

$$R_{ab} = R_{ab}(0) \tag{2.43}$$

The (periodic) autocorrelation sequence of the sequence $\{c_k\}$ is also given by

$$R_{cc}(i) = \frac{1}{L} \cdot \sum_{k=0}^{L-1} c_k \cdot c_{|k+i|_L} . \tag{2.44}$$

We can easily show that the autocorrelation sequence of an *m*-sequence is

$$R_{cc}(i) = \begin{cases} 1 & i = 0, \\ -1/L & i \neq 0. \end{cases} \tag{2.45}$$

Such an autocorrelation function is similar to that of a delta-correlated white noise process. Therefore, owing to this 'noise-like' behavior, such sequences are also referred to as *Pseudo-Noise* (PN).

In the spreading arrangement (2.30) discussed above the spreading code symbols c_k are real-valued, and therefore such an SS scheme is referred to as *Real Spreading* (RS). In particular, if the data $\tilde{d}_k$ are complex-valued the SS format in (2.30) is called *Quadrature Real Spreading* (Q-RS).

This scheme is probably the most intuitive, but it is not the only one. Further spreading schemes involving different spreading codes for the I/Q components are used in the practice. For instance, if we let (see Figure 2-7)

$$\tilde{s}^{(SS)}(t) = A \cdot \sum_{k=-\infty}^{\infty} \left(d_{I,\{k\}_M} \cdot c_{I,|k|_L} + j d_{Q,\{k\}_M} \cdot c_{Q,|k|_L} \right) \cdot g_T(t - kT_c) \tag{2.46}$$

we have two information bearing symbol streams $\{d_{I,\{k\}_M}\}$, $\{d_{Q,\{k\}_M}\}$ and two distinct spreading sequences $\{c_{I,|k|_L}\}$, $\{c_{Q,|k|_L}\}$. As a consequence the complex spreading scheme (2.46) can be seen as a combination of two RS schemes operating independently, the one for the I-, and the other for the Q-finger, respectively. This is called *dual Real Spreading* (d-RS) since it uses two different code sequences to independently spread two different data streams. By replicating the same data stream on both the I- and Q-finger of (2.46), we can derive a particular case of spreading,. i.e.,

$$d_{I,\{k\}_M} = d_{Q,\{k\}_M} = d_{\{k\}_M} \tag{2.47}$$

with $d_{\{k\}_M}$ real-valued, thus obtaining

$$\tilde{s}^{(SS)}(t) = A \cdot \sum_{k=-\infty}^{\infty} d_{\{k\}_M} \cdot \left(c_{I,|k|_L} + j c_{Q,|k|_L} \right) \cdot g_T(t - kT_c). \tag{2.48}$$

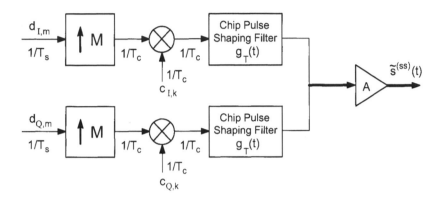

Figure 2-7. Baseband block diagram of a DS/SS transmitter
employing two spreading sequences.

The latter SS format is named *Complex Spreading* (CS) since it uses a complex-valued sequence to spread a stream of real-valued symbols. A performance comparison among the different spreading schemes outlined above will be presented in a subsequent Section.

Now, let us focus our attention on the *detection* of a DS/SS signal. The simplified architecture of an I/Q receiver for a DS/SS signal is shown in Fig 2-8. The received signal undergoes I/Q baseband conversion by means of a front end such as the one depicted in Figure 2-3. Assuming for simplicity a DS/SS like in (2.30), the signal after baseband conversion is (perfect carrier recovery)

$$\tilde{r}(t) = A \cdot \sum_{k=-\infty}^{\infty} \tilde{d}_{\{k\}_M} \cdot c_{|k|_L} \cdot g_T(t - kT_c) + \tilde{w}(t), \tag{2.49}$$

where $\tilde{w}(t)$ is Additive White Gaussian Noise (AWGN) representing the complex envelope of the channel noise affecting the modulated signal. The noise process can be expressed as

$$\tilde{w}(t) \overset{\Delta}{=} w_I(t) + j w_Q(t), \tag{2.50}$$

where $w_I(t)$ and $w_Q(t)$ are two baseband signals representing the in phase and quadrature components, respectively, of the AWGN process. Such components are independent, zero mean, white Gaussian random processes with two-sided PSD

$$\mathcal{P}_{w_I}(f) = \mathcal{P}_{w_Q}(f) = N_0.$$
(2.51)

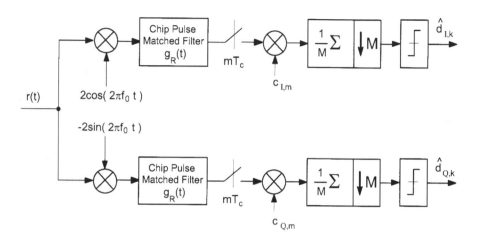

Figure 2-8. Block diagram of a DS/SS receiver.

After baseband conversion each component of the complex-valued received signal (2.49) is passed through a filter matched to the chip pulse shape (Chip Matched Filter, CMF). Assuming Nyquist's SRRC chip pulse shaping, the I and Q receive filters have frequency response given by

$$G_R(f) = \sqrt{G_{Nyq}(f)/T_c}$$
(2.52)

The filtered signal is then

$$\tilde{y}(t) = A \cdot \sum_{k=-\infty}^{\infty} \tilde{d}_{\{k\}_M} \cdot c_{|k|_L} \cdot g_{Nyq}(t - kT_c) + \tilde{n}(t),$$
(2.53)

where we have introduced the (chip time) Nyquist's RC pulse (see (2.52) and (2.27))

$$g_{Nyq}(t) = g_T(t) \otimes g_R(t)$$
(2.54)

and we have defined the filtered zero mean Gaussian noise process

$$\tilde{n}(t) = \tilde{w}(t) \otimes g_R(t) \qquad (2.55)$$

with $\otimes$ denoting convolution. The signal output by the filter is then sampled at chip rate. In the case of perfect chip clock recovery, the sampling instants are $t_m = mT_c$ and the resulting chip rate digital signal is

$$\tilde{y}_m = \tilde{y}(t_m) = A \cdot \sum_{k=-\infty}^{\infty} \tilde{d}_{\{k\}_M} \cdot c_{|k|_L} \cdot g_{Nyq}\left[(m-k)T_c\right] + \tilde{n}_m. \qquad (2.56)$$

The noise component in (2.56) is defined as

$$\tilde{n}_m = n_{I,m} + jn_{Q,m} = \tilde{n}(t_m), \qquad (2.57)$$

whose I/Q components ($n_{I,m}, n_{Q,m}$) are independent identically distributed zero mean Gaussian random variables with variance $\sigma_{n_I}^2 = \sigma_{n_Q}^2$ given by (see (2.51))

$$\sigma_{n_I}^2 = \int_{-\infty}^{\infty} P_{w_I}(f)|G_R(f)|^2 \, df = N_0 \int_{-\infty}^{\infty} |G_R(f)|^2 \, df = \frac{N_0}{T_c}. \qquad (2.58)$$

Recalling the features of the Nyquist's RC pulse (2.25), we remark that $g_{Nyq}[(m-k)T_c]$ is zero for $m \neq k$ and is 1 when $m = k$ (Nyquist-1 pulse) and we are left with

$$L \gg 1 \qquad (2.59)$$

The receiver generates a replica of the spreading sequence used by the transmitter (the co called *replica code*), and executes the product between the chip rate samples $\tilde{y}_m$ and the chips of such local code replica to 're-move' the code from the received signal. In practice this is equivalent to the operation of *despreading* carried out by conventional analog SS receivers. After multiplication (despreading) the samples are accumulated over an M-symbol long window and finally downsampled at symbol rate R_s to yield the decision strobes $\tilde{z}_k$ for data detection

$$\tilde{z}_k = \frac{1}{M} \cdot \sum_{m=kM}^{kM+M-1} \tilde{y}_m \cdot c_{|m|_L}. \qquad (2.60)$$

Equation (2.60) actually describes a digital integrator over the symbol time which performs symbol matched filtering. Substituting (2.59) in (2.60), we have

$$\tilde{z}_k = \frac{A}{M} \cdot \sum_{m=kM}^{kM+M-1} \tilde{d}_{\{m\}_M} \cdot c_{|m|_L} \cdot c_{|m|_L} + \tilde{v}_k .$$ (2.61)

In (2.61) we have a noise term

$$\tilde{v}_k = v_{I,k} + j v_{Q,k} = \frac{1}{M} \cdot \sum_{m=kM}^{kM+M-1} \tilde{n}_m \cdot c_{|m|_L}$$ (2.62)

whose I/Q components are independent, identically distributed, zero mean Gaussian random variables with variance

$$\sigma_{v_I}^2 = \sigma_{v_Q}^2 = \frac{1}{M^2} \cdot M \sigma_{n_I}^2 = \frac{N_0}{MT_c} = \frac{N_0}{T_s} .$$ (2.63)

Recalling the definitions in (2.31), we observe that, for $kM \leq m \leq kM + M - 1$ we have $\{m\}_M = k$, thus

$$\tilde{z}_k = A \cdot \tilde{d}_k \frac{1}{M} \sum_{m=kM}^{kM+M-1} c_{|m|_L} \cdot c_{|m|_L} + \tilde{v}_k$$ (2.64)

or

$$\tilde{z}_k = A \cdot \tilde{d}_k \frac{1}{M} \sum_{m=kM}^{kM+M-1} c^2_{|m|_L} + \tilde{v}_k .$$ (2.65)

But for binary spreading codes

$$c^2_{|m|_L} = 1,$$ (2.66)

and we eventually obtain the following decision variable for data detection

$$\tilde{z}_k = A \cdot \tilde{d}_k + \tilde{v}_k .$$ (2.67)

Such a decision strobe is eventually passed to the final detector, which is just a slicer to regenerate the transmitted digital data stream if no channel coding is used. The cascade of multiplication, accumulation, and downsampling produces a 'sufficient statistics' which is the same as in a conventional matched filter narrowband receiver. Actually, from (2.60) we see that the cascade of despreading and accumulation can be seen also as the computation of a *correlation* between the incoming signal (2.56) and the local code replica, and the downsampling is aimed at taking the maximum of the correlation. This is why the detector in Figure 2-8 is also called the *Correlation Receiver* (CR). We remark again that the decision strobe in (2.67) is exactly the same we would obtain in the case of a conventional narrowband transmission of complex-valued symbols $\tilde{d}_k$ over AWGN channel and using a matched filter receiver. Therefore the spreading and despreading operations, carried out in the transmitter and receiver, respectively, are completely transparent to the final user.

The Bit Error Rate (BER) $P(e)$ of a DS/SS transmission over an AWGN channel is therefore coincident with that of a conventional narrowband transmission employing the same modulation format. For instance, in the case of a Binary PSK (BPSK) or a Quadrature PSK (QPSK) we obtain (see Figure 2-9)

$$P(e) = Q\left(\sqrt{\frac{2E_b}{N_0}}\right), \tag{2.68}$$

where $E_b = P_s T_b$ is the average received energy at RF per information bit and $Q(x)$ is the Gauss' integral function

$$Q(x) = \int_x^\infty \frac{1}{\sqrt{2\pi}} e^{-y^2/2} \, dy. \tag{2.69}$$

According to the theoretical description outlined above, the detection of a DS/SS signal can be accomplished by using the receiver architecture shown in Figure 2-8, whose baseband equivalent block diagram is depicted in Figure 2-10. It is also apparent that DS/SS communications require more complex receivers with respect to conventional narrowband signaling [De98b]. Suffice it to say that sampling at the filter output must be performed at chip (instead of symbol) rate, and this requires much faster signal processing capabilities. Furthermore, timing recovery must be carried out with respect to the chip (instead of symbol) interval, and this is much more demanding in terms of accuracy of synchronization. Finally, as can be inferred from (2.61), a proper alignment between the received signal and the local replica of the spreading sequence is mandatory in order to successfully perform signal despreading through a perfectly aligned correlation.

From the considerations made above, it is evident that the most peculiar and crucial function which the DS/SS receiver has to cope with is timing recovery. The basic difference between the function of symbol timing recovery in a conventional modem for narrowband signals and code alignment in a wideband SS receiver lies in a fundamental difference in the statistical properties of the data bearing signal. In narrowband modulation the data signal bears an intrinsic statistical regularity on a symbol interval T_s that is, properly speaking, it is *cyclostationary* with period T_s. Clock recovery is to be carried out with an accuracy of some hundredths of a T_s, and is not particularly troublesome. Owing to the presence of the spreading code, the DS/SS signal is cyclostationary with period LT_c (in a short code arrangement), but the receiver has to derive a timing estimate with an accuracy comparable to *a tenth of the chip interval T_c* to perform correlation and avoid *Inter-Chip Interference* (ICI).

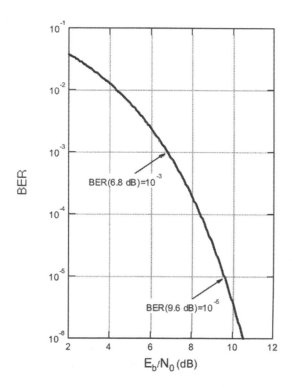

Figure 2-9. BER of a matched-filter receiver for BPSK / QPSK transmission over the Gaussian channel.

This simple discussion suggests that timing estimation becomes more and more involved as L gets large (long codes). Unfortunately, in practical applications of DS/SS transmissions we always have $L \gg 1$ even for short codes (typically $L \geq 31$), so that the problem of signal timing recovery with a sufficient accuracy is much more challenging for wideband DS/SS signals than for narrowband modulation, and is usually split in the two phases of coarse acquisition and fine tracking. The first is activated during receiver startup, when the DS/SS demodulator has to find out whether the intended user is transmitting, and, in the case in which he/she actually is, coarsely estimate the signal delay to initiate fine chip time tracking and data detection. Code tracking is started upon completion of the acquisition phase and aims at locating the optimum sampling instant of the chip rate signal to provide ICI-free samples (such as (2.59)) to the subsequent digital signal processing functions.

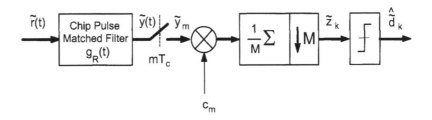

Figure 2-10. Baseband equivalent of a DS/SS receiver.

After examining the main functions for signal detection, we present some introductory considerations about the practical implementation of a DS/SS receiver. In this respect Figure 2-11 shows a scheme of a DS/SS receiver highlighting also the different signal synchronization functions (carrier frequency/phase and timing) which often represent the real crux of good modem design. We have denoted by $\Delta\hat{f}$, $\hat{\theta}$, and $\hat{\tau}$ the estimates of the carrier frequency offset, phase offset, and chip timing error, respectively, relevant to the useful signal. As already discussed (see Figure 2-8), the baseband I/Q components of $r(t)$ are derived via a baseband I/Q converter as the one in Figure 2-3. Such a converter is usually implemented at IF in double conversion receivers or directly at RF in low cost, low power receivers (this is the case, for instance, for mobile phones).

The basic architecture of Figure 2-11 can be entirely implemented via DSP components by performing Analog to Digital Conversion (ADC) as early as possible, at times directly on the IF (intermediate frequency) signal provided at the output of the RF to IF front end conversion stage in the receiver. In so doing, the baseband received signal $\tilde{r}(t)$ in Figure 2-11 is actually a sampled digital signal, carrier recovery and chip matched filtering are digital, and the 'sampler' is just a decimator/interpolator that changes the clock rate of the digital signal. The ADC conversion rate of $\tilde{r}(t)$ is, in fact, invariably faster than the chip rate to perform chip matched filtering with no aliasing problems. We shall say more about the digital architecture of the DS/SS receiver in Chapter 3 when dealing specifically with the MUSIC demodulator.

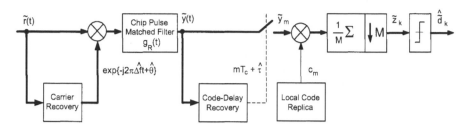

Figure 2-11. Architecture of a receiver for DS/SS signals, including synchronization units.

3. CODE DIVISION MULTIPLEXING AND MULTIPLE ACCESS

In the DS/SS schemes discussed above the data stream generated by an information source is transmitted over a wide frequency spectrum using one (or two) spreading code(s). Starting from this consideration we can devise an access system allowing *multiple* users to share a common channel transmitting their data in DS/SS format. This can be achieved by assigning each user a different spreading code and allowing all the signals *simultaneously* access, in DS/SS mode, the *same frequency spectrum*. All the user signals are therefore transmitted at the *same time* and over the *same frequency band*, but they can nevertheless be identified thanks to the particular spreading code used, which is different from one user to another (the so called *signature code*). The users are separated in the *code domain*, instead of time or frequency domain, as in conventional *Time* or *Frequency Division Multiple Access*, respectively (TDMA, FDMA). Such a multiple access technique, based on DS/SS transmission, is therefore called *Code Division Multiple Access* (CDMA) and the spreading sequence identifying each user is also referred to as *signature*. The N user signals in DS/SS format can be obtained from a set of N tributary channels made available to a single transmitting unit which performs spectrum spreading of each of them, followed by Code Division Multiplexing (CDM). Alternatively, the DS/SS signals can be originated by N spatially separated terminals, and in this latter case code division multiplexing occurs at the receiver antenna.

Let us focus our attention on the detection of a DS/SS signal in the case of a *multiuser* CDMA system in which N users are concurrently active. For the sake of simplicity, we refer once again to the simplified signal model

(2.30). We will start considering the case a CDMA multiuser communication in which all of the spreading sequences of the different users are *synchronous*, i.e., the start epoch is exactly the same for each code. We will refer to this arrangement as *Synchronous* CDMA (S-CDMA). This is the case of a CDMA signal originated from a single transmitter, i.e., from a base station (or satellite) to a group of mobile receivers. We will therefore address such a scenario as *single-cell*. The received signal, after baseband conversion and under the hypothesis of perfect carrier recovery, can be written as

$$\tilde{r}(t) = \sum_{i=1}^{N} A^{(i)} \cdot \sum_{k=-\infty}^{\infty} \tilde{d}_{\{k\}_M}^{(i)} \cdot c_{|k|_L}^{(i)} \cdot g_T(t - kT_c) + \tilde{w}(t), \qquad (2.70)$$

with the same definitions as in the single-user case described by (2.49), where for each user's channel we have defined the amplitude coefficient (see (2.11))

$$A^{(i)} = \sqrt{2P_{s_i} / A_d^2} . \qquad (2.71)$$

Notice that, in order to take the multiple users accessing the RF spectrum into account we have introduced the superscript $^{(i)}$ which identifies the amplitude, data, and code chips of the generic i th user. The generation of the aggregate code division multiplexed signal (2.70) is conceptually depicted in Figure 2-12

Assuming now, without loss of generality, that the receiver intends to detect the data transmitted by user 1, we can re-write (2.70) as

$$\tilde{r}(t) = A^{(1)} \cdot \sum_{k=-\infty}^{\infty} \tilde{d}_{\{k\}_M}^{(1)} \cdot c_{|k|_L}^{(1)} \cdot g_T(t - kT_c)$$

$$+ \sum_{i=2}^{N} A^{(i)} \cdot \sum_{k=-\infty}^{\infty} \tilde{d}_{\{k\}_M}^{(i)} \cdot c_{|k|_L}^{(i)} \cdot g_T(t - kT_c) + \tilde{w}(t), \qquad (2.72)$$

where the first term at the right hand side is the useful signal to be detected, while the second one, denoted in a more compact form as

$$\tilde{b}(t) = \sum_{i=2}^{N} A^{(i)} \cdot \sum_{k=-\infty}^{\infty} \tilde{d}_{\{k\}_M}^{(i)} \cdot c_{|k|_L}^{(i)} \cdot g_T(t - kT_c) \qquad (2.73)$$

is an additional component owed to multiple access. In a conventional correlation receiver, the received signal (2.72), is passed through the chip matched filter and sampled at chip rate yielding the samples $\tilde{y}_m$ (see (2.59))

$$\tilde{y}_m = A^{(1)} \cdot \tilde{d}^{(1)}_{\{m\}_M} \cdot c^{(1)}_{|m|_L} + \sum_{i=2}^{N} A^{(i)} \cdot \tilde{d}^{(i)}_{\{m\}_M} \cdot c^{(i)}_{|m|_L} + \tilde{n}_m$$

$$= A^{(1)} \cdot \tilde{d}^{(1)}_{\{m\}_M} \cdot c^{(1)}_{|m|_L} + \tilde{b}_m + \tilde{n}_m , \qquad (2.74)$$

where $\tilde{b}_m = \tilde{b}(t_m)$ is given by

$$\tilde{b}_m = b_{I,m} + jb_{Q,m} = \sum_{i=2}^{N} A^{(i)} \cdot \tilde{d}^{(i)}_{\{m\}_M} \cdot c^{(i)}_{|m|_L} . \qquad (2.75)$$

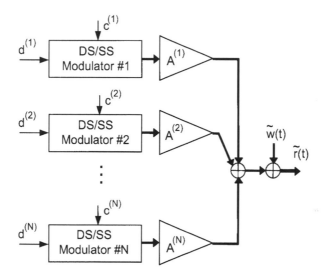

Figure 2-12. Generation of a multiuser S-CDMA signal.

The I/Q components $b_{I,m}$ and $b_{Q,m}$ in (2.75) are independent, identically distributed, zero mean random variables whose variance is

$$\sigma_{b_I}^2 = \sigma_{b_Q}^2 = \mathrm{E}\{b_{I,m}^2\} . \qquad (2.76)$$

The sampled signal (2.74) undergoes correlation (despreading / accumulation) with the signature code of user 1 as follows

$$\tilde{z}_k^{(1)} = \frac{1}{M} \cdot \sum_{m=kM}^{kM+M-1} \tilde{y}_m \cdot c^{(1)}_{|m|_L} . \qquad (2.77)$$

After some algebra we find (see (2.61)–(2.64))

$$\tilde{z}_k^{(1)} = \frac{A^{(1)} \cdot \tilde{d}_k^{(1)}}{M} \cdot \sum_{m=kM}^{kM+M-1} c_{|m|_L}^{(1)} \cdot c_{|m|_L}^{(1)}$$

$$+ \sum_{i=2}^{N} \frac{A^{(i)} \cdot \tilde{d}_k^{(i)}}{M} \cdot \sum_{m=kM}^{kM+M-1} c_{|m|_L}^{(i)} \cdot c_{|m|_L}^{(1)} + \tilde{v}_k \qquad (2.78)$$

and

$$\tilde{z}_k^{(1)} = A^{(1)} \cdot \tilde{d}_k^{(1)} \cdot \chi_{c^{(1)}c^{(1)}}(k) + \sum_{i=2}^{N} A^{(i)} \cdot \tilde{d}_k^{(i)} \cdot \chi_{c^{(i)}c^{(1)}}(k) + \tilde{v}_k, \qquad (2.79)$$

where we have defined the following partial auto- and cross-correlations

$$\chi_{c^{(1)}c^{(1)}}(k) = \frac{1}{M} \cdot \sum_{m=kM}^{kM+M-1} c_{|m|_L}^{(1)} \cdot c_{|m|_L}^{(1)} = 1, \qquad (2.80)$$

$$\chi_{c^{(i)}c^{(1)}}(k) = \frac{1}{M} \cdot \sum_{m=kM}^{kM+M-1} c_{|m|_L}^{(i)} \cdot c_{|m|_L}^{(1)}. \qquad (2.81)$$

The decision strobe is eventually passed to the final detector which regenerates the transmitted digital data stream of the user 1 (the desired, 'singing' user). From (2.79) it is apparent that the decision strobe $\tilde{z}_k^{(1)}$ is composed of three terms: i) the useful datum (first term); ii) Gaussian noise (third term); and iii) an additional term arising form the concurrent presence of multiple users and called *Multiple Access Interference* (MAI). In particular, the MAI term can be expressed as

$$\tilde{\beta}_k = \beta_{I,k} + j\beta_{Q,k} = \sum_{i=2}^{N} A^{(i)} \cdot \tilde{d}_k^{(i)} \cdot \chi_{c^{(i)}c^{(1)}}(k), \qquad (2.82)$$

or equivalently, according to definition (2.75), as

$$\tilde{\beta}_k = \beta_{I,k} + j\beta_{Q,k} = \frac{1}{M} \cdot \sum_{m=kM}^{kM+M-1} \tilde{b}_m \cdot c_{|m|_L}^{(1)} \qquad (2.83)$$

The I/Q components $\beta_{I,k}$ and $\beta_{Q,k}$ are independent, identically distributed, zero mean random variables whose variance can be put in a form similar to (2.63)

$$\sigma_{\beta_I}^2 = \sigma_{\beta_Q}^2 = E\{\beta_{I,m}^2\} = I_0 / T_s \tag{2.84}$$

where we have introduced an equivalent PSD I_0 of the MAI term, assuming implicitly that it can be considered *flat* (white) over the whole signal spectrum. Now re-cast (2.79) into the form

$$\tilde{z}_k^{(1)} = A^{(1)} \cdot \tilde{d}_k^{(1)} + \tilde{\beta}_k + \tilde{v}_k . \tag{2.85}$$

Under certain hypotheses which we will discuss in a little while, the MAI contribution can be modeled as an additional (white) Gaussian noise (independent of $\tilde{v}_k$). Therefore the BER performance of the DS/SS signal can be analytically derived simply by assuming an equivalent noise term $\tilde{v}_k' = \tilde{\beta}_k + \tilde{v}_k$ with a total, equivalent PSD given by

$$N_0' = N_0 + I_0 , \tag{2.86}$$

and the decision strobe becomes equivalent to that in (2.67), which refers to a pure AWGN channel

$$\tilde{z}_k^{(1)} = A^{(1)} \cdot \tilde{d}_k^{(1)} + \tilde{v}_k' . \tag{2.87}$$

Consequently the BER for QPSK modulation in the presence of Gaussian MAI, can be obtained by a simple modification of expression (2.68)

$$P(e) = Q\left(\sqrt{\frac{2E_b}{N_0'}}\right) = Q\left(\sqrt{\frac{2E_b}{N_0 + I_0}}\right) . \tag{2.88}$$

If very long pseudo-random (i.e., noise-like) spreading sequences are used then the chips $c_{|m|_L}^{(i)}$ of each user code can be approximately modeled as independent random variables belonging to the alphabet $\{-1,+1\}$. Also, the chips of different users can be modeled as uncorrelated random variables. It follows that if $N \gg 1$ and if all of the signal powers are (almost) equal (i.e., $P_{s_i} = P_s$, $\forall i$), then the power of the MAI is $P_{\text{MAI}} = (N-1)P_s$ and by virtue of the central limit theorem, we can model the MAI components $\beta_{I,k}$ and $\beta_{Q,k}$ at the detector input as independent identically distributed zero mean Gaussian random variables with variance (see (2.76) and (2.83))

$$\sigma_{\beta_I}^2 = \sigma_{\beta_Q}^2 = \frac{I_0}{T_s} = \frac{P_{\text{MAI}}/(1/T_c)}{MT_c} = \frac{(N-1)P_s}{M} . \tag{2.89}$$

This situation is actually experienced, for instance, in a CDMA system with accurate power control, so that all the users signals are received at (almost) the same power level. Under this hypothesis the PSD of the MAI is

$$I_0 = T_s \frac{(N-1)P_s}{M} = (N-1)P_sT_c = (N-1)E_c = \frac{(N-1)}{G_p}E_b, \tag{2.90}$$

where $E_c = P_sT_c$ represents the average energy at RF per chip, and according to (2.40) we have set $E_c = E_b / G_p$. The BER (2.88) becomes then

$$P(e) = Q\left(\sqrt{\frac{2E_b}{N_0 + (N-1)E_c}}\right) \tag{2.91}$$

and with some manipulations we obtain for QPSK

$$P(e) = Q\left(\sqrt{\frac{2E_b}{N_0}} \cdot \frac{1}{\sqrt{1 + \frac{(N-1)}{M}\frac{2E_b}{N_0}}}\right). \tag{2.92}$$

From the expressions above it turns out that the MAI degrades the BER performance. In particular, the degradation increases with the number of interfering channels and decreases for large processing gains. Notice also that, in the particular case $N = 1$ (2.92) collapses to the conventional BER expression relevant to (narrowband) QPSK modulation over AWGN channel and matched filter detection.

However, we must remark that in the more general case of CDMA transmissions with MAI ($N > 1$) (2.92) is accurate only under certain conditions. In particular, the assumption of uncorrelated binary random variables for the code chips is valid only when the signature codes are 'long' in the sense of Section 2. As is apparent from (2.82), the amount of MAI is in reality determined by the cross-correlation properties between the useful signal and the interferers. Therefore, in order to derive a more accurate analytical expression for the BER the particular type of spreading codes and the relevant correlation properties must be accounted for. In order to simplify the analytical description, from now on we shall focus on the case of short spreading codes, i.e., $M = n \cdot L$. Recalling (2.42), the cross-correlation (2.81) is now

$$\chi_{c^{(i)}c^{(1)}}(k) = \frac{1}{nL} \cdot \sum_{m=knL}^{knL+nL-1} c_{|m|_L}^{(i)} \cdot c_{|m|_L}^{(1)} = \frac{1}{nL} \cdot n \cdot \sum_{m=knL}^{knL+L-1} c_{|m|_L}^{(i)} \cdot c_{|m|_L}^{(1)} = R_{c^{(i)}c^{(1)}} \cdot \quad (2.93)$$

The variance of the I/Q components of the MAI samples $\tilde{\beta}_m$ must be re-written by resorting to (2.82), yielding

$$\sigma_{\beta_I}^2 = \sigma_{\beta_Q}^2 = \frac{I_0}{T_s} = \sum_{i=2}^{N} P_{s_i} R_{c^{(i)}c^{(1)}}^2 \,, \quad (2.94)$$

and the PSD of the MAI contribution to the total noise in (2.86) becomes

$$I_0 = T_s \sum_{i=2}^{N} P_{s_i} R_{c^{(i)}c^{(1)}}^2 \cdot \quad (2.95)$$

In the case of equi-powered users we obtain

$$I_0 = P_s T_s \sum_{i=2}^{N} R_{c^{(i)}c^{(1)}}^2 = E_s \sum_{i=2}^{N} R_{c^{(i)}c^{(1)}}^2 \,, \quad (2.96)$$

where $E_s = P_s T_s$ represents the average energy at RF per modulation symbol. Since, for QPSK, $R_s = R_b / 2$, we have $E_s = 2E_b$ and therefore

$$I_0 = 2E_b \sum_{i=2}^{N} R_{c^{(i)}c^{(1)}}^2 \quad (2.97)$$

By retaining the assumption of a Gaussian distribution of the MAI, which holds true in the case of large spreading factors and large number of users, the BER is now

$$P(e) = Q\left(\sqrt{\frac{2E_b}{N_0}} \cdot \frac{1}{\sqrt{1 + \sum_{i=2}^{N} R_{c^{(i)}c^{(1)}}^2 \frac{2E_b}{N_0}}} \right) \cdot \quad (2.98)$$

From the expression above we can conclude that in order to limit the detrimental effect of MAI on BER performance the spreading sequences must be chosen so as to exhibit the lowest possible cross-correlation level. In the case of maximal length sequences with $L \gg 1$, the cross-correlation is well approximated by [Sar80]

$$R_{c^{(i)}c^{(1)}} \cong 1/\sqrt{L} \tag{2.99}$$

thus, recalling that $M = n \cdot L$, we obtain

$$P(e) = Q\left(\sqrt{\frac{2E_b}{N_0}} \cdot \frac{1}{\sqrt{1 + \dfrac{n \cdot (N-1)}{M} \dfrac{2E_b}{N_0}}}\right), \tag{2.100}$$

which for $n = 1$ coincides with the BER expression (2.92) previously de-rived for the white Gaussian MAI model. Actually (2.99) represents the RMS value of the cross-correlation between two L-period maximal length sequences taken over all the possible relative phase shifts. However, it is found that, in spite of their many appealing features, m-sequences are not convenient for CDMA. First, for a given m there exists only a limited num-ber of sequences available for user identification in a CDMA system [Din98]. Also, the cross-correlation properties of m-sequences are not opti-mal, so they result in significant levels of MAI.

The MAI term $\tilde{\beta}_k$ in (2.85) can be canceled by using *orthogonal* codes such as $R_{c^{(i)}c^{(j)}} = 0$ ($i \neq j$). A popular set of orthogonal spreading codes is represented by the *Walsh–Hadamard* (WH) sequences [Ahm75], [Din98] which have period $L = 2^m$ and are obtained taking the rows (or the columns) of the $L \times L$ matrix $\mathbf{H}_m$ recursively defined as follows

$$\mathbf{H}_m = \begin{bmatrix} \mathbf{H}_{m-1} & \mathbf{H}_{m-1} \\ \mathbf{H}_{m-1} & -\mathbf{H}_{m-1} \end{bmatrix}, \quad \mathbf{H}_1 = \begin{bmatrix} +1 & +1 \\ +1 & -1 \end{bmatrix} \tag{2.101}$$

where $-\mathbf{H}_m$ means the complement (i.e., the sign inversion) of each element of the matrix $\mathbf{H}_m$. From (2.101) it is apparent that for a given period L the WH set is composed of L sequences. Thanks to orthogonality the BER per-formance for an *Orthogonal* CDMA (O-CDMA) system is obtained by re-moving the MAI contribution in (2.98), which gives the conventional ex-pression for narrowband BPSK/QPSK modulation (2.68).

Despite such an appealing feature, it must be noticed that the WH se-quences exhibit very poor off zero auto- and cross-correlation properties making difficult initial code acquisition and user recognition by the receiver. For this reason, in practical applications *pure* orthogonal codes such as the WH sequences must be used overlaid by a PN sequence [Fon96], [Din98]. According to this approach the resulting *composite code* is therefore the *su-perposition* of two codes, i.e., an orthogonal WH code $\{c_k^{(i)}|_{\text{WH}}\} \in \{\pm 1\}$ for user identification (the so called *traffic* or *channelization code*) plus an *overlay* PN sequence $\{c_k|_{PN}\} \in \{\pm 1\}$, common to all the users within the same cell (or satellite beam), as follows

$$c_k^{(i)} \triangleq c_k|_{PN} \cdot c_k^{(i)}|_{WH} \ . \tag{2.102}$$

The cell- (or beam-) unique overlay code yields a twofold benefit: first, it represents a sort of cell (or beam) identifier and second, it performs a 'randomization' of the user signature that is helpful in reducing unwanted off zero auto- and cross-correlation peaks. For this reason the overlay code is also called *scrambling code*. It is immediate to observe also that orthogonality between any pair of composite sequences is preserved, i.e., $R_{c^{(i)}c^{(j)}} = 0$ ($i \neq j$) still holds true. Finally, notice that if we want to obtain a composite code having exactly the same period $L = 2^m$ as the original WH sequence we must select an overlay maximal length sequence having period $2^m - 1$, and properly extend it by inserting a '+1' chip into its longest run. Such a modified sequence is called *Extended* PN (E-PN). The use of orthogonal codes (either simple or composite) cancels the MAI term $\tilde{\beta}_k$ out of (2.79), yielding the very same decision strobe as in the single-user case (2.67).

Other sets of codes widely used as spreading sequences in practical CDMA systems are the *quasi-orthogonal* ones. These codes have non-null (yet small) cross-correlation ($R_{c^{(i)}c^{(j)}} \neq 0$, for $i \neq j$), but exhibit less critical off zero correlation performance. For instance, by a proper combination of two selected PN sequences with period $L = 2^m - 1$, we obtain the *Gold* sequences [Gol67], [Gol68], [Sar80], [Din98], which have period L, cross-correlation $R_{c^{(i)}c^{(1)}} = -1/L$ and small 3-valued off zero cross-correlation. It is also found that the number of Gold codes having period L is $L + 2$ (apart from the particular case $L = 255$ which admits only $L + 1 = 256$ codes) [DeG91]. From (2.98), and recalling that $M = n \cdot L$, the BER performance for a *Quasi-Orthogonal* CDMA (QO-CDMA) system employing QPSK modulation, becomes

$$P(e) = Q\left(\sqrt{\frac{2E_b}{N_0}} \cdot \frac{1}{\sqrt{1 + \dfrac{n^2 \cdot (N-1)}{M^2} \dfrac{2E_b}{N_0}}} \right) \tag{2.103}$$

and, from (2.97), the PSD of the equivalent noise owed to MAI is

$$I_0 = \frac{n^2 \cdot (N-1)}{M^2} 2E_b \ .$$

(2.104)

Similarly to the PN, Gold sequences can also be extended by proper insertion of an additional chip into one sequence period in order to obtain a set of quasi-orthogonal codes with repetition period $L = 2^m$ which is called *Extended Gold* (E-GOLD). Another set of quasi-orthogonal codes is the *Kasami* set [Kas68], [Sar80], [Din98]. The first step to obtain a Kasami sequence is decimation of an *m*-sequence, with *m* even, by a factor $s = 2^{m/2} + 1$ (thus obtaining a further *m*-sequence with period $2^{m/2} - 1$ [Sar80]), and extension by repetition (*s* times) of the decimated sequence up to the original length. The set is then constructed by collecting all of the sequences obtained by addition of any cyclical shift of the decimated/extended sequence to the original *m*-sequence, and including the original sequence as well. The total number of elements (sequences) in the set is thus $2^{m/2}$. The cross-correlation sequence for two Kasami sequences takes on the three values: -1, $-s$, and $s - 2$.

The *large* set of Kasami sequences consists of sequences of period $2^{m/2} - 1$, with *m* even, and contains both a set of Gold (or Gold-like) sequences and the small set of Kasami sequences as subsets. To obtain the Kasami large set, we start with two equal length *m*-sequences *y* and *z* both obtained after decimation/repetition of a 'mother' longer *m*-sequence *x* as above. We then take all of the sequences obtained by adding *x*, *y*, and *z* with any possible (cyclical) shifts of *y* and *z*, for a total number of $2^{m/2}(2^m + 1)$ sequences if $|m|_4 = 2$, and $2^{m/2}(2^m + 1) - 1$ if $|m|_4 = 0$. The auto- and cross-correlation sequences are limited to 5 particular values we will not specify here (more details can be found in Kasami's seminal paper [Kas68], in the extensive investigation about codes correlation properties by Sarwate and Pursley [Sar80] and in the survey on spreading codes for DS-CDMA by Dinan and Jabbari [Din98]).

Let us now compare, in terms of capacity, the spreading arrangements previously discussed in Section 2. We assume a CDMA system with N active users, each transmitting at a bit rate R_b, and employing short spreading codes with period L and spreading factor $M = n \cdot L$. Considering, for the sake of simplicity, a set of WH codes we then have that L sequences are available to the users. System capacity performance is expressed in terms of spectral efficiency, defined as

$$\eta = \frac{N \cdot R_b}{B^{(SS)}} \quad \text{(bit/sec)/Hz,}$$

(2.105)

where the SS bandwidth is $B^{(SS)} = R_c$ (Nyquist bandwidth). Table 2-1 compares the different spreading arrangements previously outlined.

Table 2-1. Comparison among spreading arrangements

Spreading Arrangement	Constellation Symbols	$B^{(ss)}$	G_p	M	N	η [bit/s/Hz]
RS	BPSK	$n \cdot L \cdot R_b$	$n \cdot L$	$n \cdot L$	L	$1/n$
d-RS	2 x BPSK	$n \cdot L \cdot R_b/2$	$n \cdot L/2$	$n \cdot L$	$L/2$	$1/n$
CS	BPSK	$n \cdot L \cdot R_b$	$n \cdot L$	$n \cdot L$	$L/2$	$1/(2n)$
Q-RS	QPSK	$n \cdot L \cdot R_b/2$	$n \cdot L/2$	$n \cdot L$	L	$2/n$

We notice that the bandwidth occupancy and the processing gain for BPSK constellations are twice as those of QPSK (or dual BPSK), while the spreading factor is the same for all of the cases. The maximum number of active users N is given by the size L of the WH spreading codes set for those schemes employing only one sequence per uses, whilst it is half the set size for those schemes assigning two different codes (one for the I and one for the Q stream) to each user. The last column presents the spectral efficiency evaluated from (2.105) and demonstrates that Q-RS is the most efficient scheme while CS is the least one.

In the case of advanced communication systems supporting different kinds of services (e.g., voice, video, data), the user bit rates can be variable from a few kbit/s up to hundreds of Mbit/s. In these cases the spreading scheme will be flexible enough to easily allocate signals with different bit rates on the same bandwidth. This can be achieved by maintaining a fixed chip rate R_c (and therefore a fixed spread spectrum bandwidth $B^{(SS)}$) and by concurrently varying the spreading factor M according to the bit rate of the signal to be transmitted. This should also be done without altering the property of mutual orthogonality outlined above. The solution to this problem is the special class of codes named *Orthogonal Variable Spreading Factor* (OSVF) codes [Ada97], [Din98]. The OVSF code set is a re-organization of the Walsh–Hadamard codes into *layers*. The codes on each layer, as is shown in Figure 2-13, have twice the length of the codes in the layer above. In addition the codes are organized in a *tree*, in which any two 'children' codes on the layer underneath a 'parent' code are generated by repetition, and repetition with sign change, respectively.

The peculiarity of the tree is that any two codes are not only orthogonal within each layer (that is just the complete set of the Walsh–Hadamard codes of the corresponding length), but they are also orthogonal *between* layers (after extension by repetition of the shorter code), provided that the shorter is not an ancestor of the longer one. As a consequence we can use the shorter code for a higher rate transmission with a smaller spreading factor, and the longer code for a lower rate transmission with a higher spreading factor (recall that the chip rate is always the same). The two codes will not give rise to any channel crosstalk (MAI).

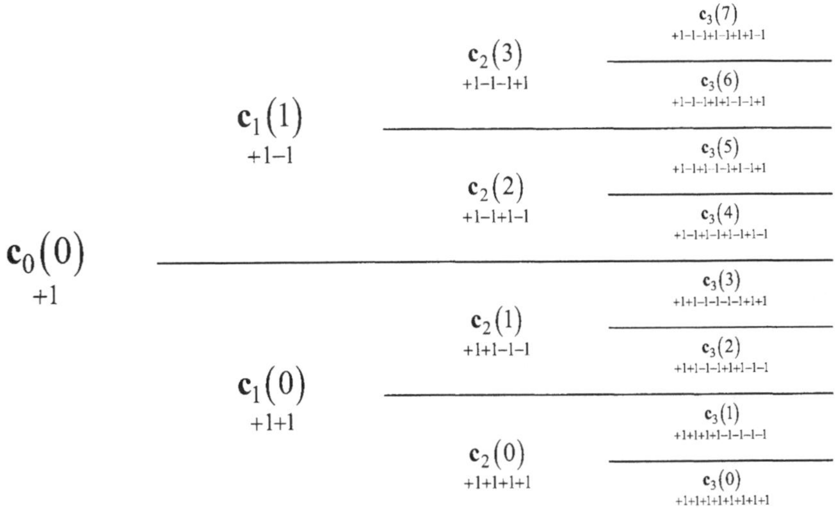

$c_3(7)$
+1-1-1+1-1+1+1-1

$c_2(3)$
+1-1-1+1

$c_3(6)$
+1-1-1+1-1-1-1+1

$c_1(1)$
+1-1

$c_3(5)$
+1-1+1-1-1+1-1+1

$c_2(2)$
+1-1+1-1

$c_3(4)$
+1-1+1-1+1-1+1-1

$c_0(0)$
+1

$c_3(3)$
+1+1-1-1-1-1+1+1

$c_2(1)$
+1+1-1-1

$c_3(2)$
+1+1-1-1+1+1-1-1

$c_1(0)$
+1+1

$c_3(1)$
+1+1+1+1-1-1-1-1

$c_2(0)$
+1+1+1+1

$c_3(0)$
+1+1+1+1+1+1+1+1

Figure 2-13. The OVSF codes tree.

In the case of the uplink of a wireless cellular system, the DS/SS signals within a single cell (or beam) are originated from sparse terminals which have different signature epochs, thus resulting in *Asynchronous* CDMA (A-CDMA). The received signal, after baseband conversion and under the hypothesis of perfect timing and carrier recovery for the user 1 (i.e., the desired one) can be obtained by modifying (2.72) as follows

$$\tilde{r}(t) = A^{(1)} \cdot \sum_{k=-\infty}^{\infty} \tilde{d}_{\{k\}_M}^{(1)} \cdot c_{|k|_L}^{(1)} \cdot g_T(t - kT_c)$$

$$+ \sum_{i=2}^{N} A^{(i)} \cdot e^{-j(2\pi \Delta f_i t + \theta_i)} \sum_{k=-\infty}^{\infty} \tilde{d}_{\{k\}_M}^{(i)} \cdot c_{|k|_L}^{(i)} \cdot g_T(t - kT_c - \tau_i) + \tilde{w}(t), \quad (2.106)$$

where τ_i, θ_i and Δf_i represent the timing, carrier phase, and carrier frequency offsets, respectively, of the ith interfering channel with respect to the useful one. Notice also that, differently from the downlink described by (2.72), the interfering signal powers $P_s^{(i)}$ in the uplink described by (2.106) are, in general, unequal. This is owed to the different propagation loss experienced by each user signal originating from a different spatial location, which can be only partially compensated for by means of a power control algorithm. Such a power unbalance is expressed by the *useful to single interferer power ratio*, defined as follows

$$C/I\big|_i = P_s^{(1)}/P_s^{(i)},$$
(2.107)

and the amplitude of the interfering signals (2.71) becomes

$$A^{(i)} = \sqrt{2P_s^{(i)}/A_{\tilde{d}}^2} = \sqrt{\frac{2P_s^{(1)}/A_{\tilde{d}}^2}{C/I\big|_i}} = \frac{1}{\sqrt{C/I\big|_i}} \cdot A^{(1)}.$$
(2.108)

In this case the CDMA signal model (2.106) is

$$\tilde{r}(t) = A^{(1)} \cdot \sum_{k=-\infty}^{\infty} \tilde{d}_{\{k\}_M}^{(1)} \cdot c_{|k|_L}^{(1)} \cdot g_T(t - kT_c)$$

$$+ A^{(1)} \cdot \sum_{i=2}^{N} \frac{1}{\sqrt{C/I\big|_i}} \cdot e^{-j(2\pi\Delta f_i t + \theta_i)} \sum_{k=-\infty}^{\infty} \tilde{d}_{\{k\}_M}^{(i)} \cdot c_{|k|_L}^{(i)} \cdot g_T(t - kT_c - \tau_i)$$

$$+ \tilde{w}(t).$$
(2.109)

In order to simplify system description and/or analysis we assume an MAI model with equi-powered interfering users, $P_I = P_s^{(i)}$, $\forall i \geq 2$. We can therefore resort to a unique *C/I* ratio, defined as

$$C/I = P_s^{(1)}/P_I.$$
(2.110)

The total amount of MAI affecting the useful signal can be expressed by means of the following useful to total interfering power ratio

$$\frac{C}{I}\bigg|_{tot} = \frac{P_s^{(1)}}{P_{MAI}}$$
(2.111)

which in the case of equi-powered interferers becomes (see (2.110))

$$\frac{C}{I}\bigg|_{tot} = \frac{P_s^{(1)}}{(N-1)\cdot P_I} = \frac{1}{N-1} \cdot \frac{C}{I}.$$
(2.112)

It is fairly apparent that asynchronous access does not allow MAI cancellation by using orthogonal sequences. In this case the decision strobe can be still expressed by (2.85) in which the statistics of the MAI term $\tilde{\beta}_k$ are determined by cross-correlation properties and other parameters of the interfering signals. In a first approximation, assuming long spreading codes and ideal power control, the BER performance of the link can be computed via the Gaussian approximation (2.92).

4.   MULTI-CELL OR MULTI-BEAM CDMA

As outlined above, multiple access can be granted with DS/SS signals by assigning different spreading codes to different users. This can be done both in the downlink of a terrestrial radio network (base to mobile) with synchronous orthogonal codes, and in the uplink (mobile to base) with asynchronous, pseudo-noise codes. But a problem arises when we run out of codes, and more users ask to access the network. With reasonable spreading factors (up to 256), the number of concurrently active channels is too low to serve a large users population like we have in a large metropolitan area, or a vast suburban area. This also applies to conventional FDMA or TDMA radio networks where the number of channels is equal to the number of carriers in the allocated bandwidth or the number of time slots in a frame, respectively. The solution to this issue lies in the notion of *cellular network with frequency re-use* as outlined in Section 2 of Chapter 1. Of course, frequency re-use has an impact on the overall network efficiency in terms of users/cell (or users/km^2) since the number of channels allocated to each cell is a fraction $1/Q$ of the overall channels allocated to the provider, where Q is the frequency re-use factor (the number of cells in a cluster). The same concept of coverage area partitioning with channel re-use applies to *multi-beam satellite networks* as those envisaged in Section 3 of Chapter 1. So to both kind of radio networks the technique of *universal frequency re-use* with CDMA signals (Section 2 in Chapter 1) is applicable as well. Focusing on the downlink, universal frequency re-use means that the *same* carrier frequency is used in each cell/beam (Figure 1-5), and that the *same* orthogonal codes set (i.e., the same channels) are used within each cell/beam on the same carrier. Of course, something has to be done to prevent neighboring users at the edge of two adjacent cells/beams and using the same WH code to heavily interfere with each other. The trick consists in using a *different scrambling code* on different cells/beams to cover the channelization WH codes as in (2.102). In a sense, we use a sort of *code re-use* technique, where *code* refers to the (orthogonal) channelization codes in each cell/beam.

Let us focus our attention on the detection of a DS/SS signal in the case of a *multi-cell* (or *-beam*) *multiuser* CDMA system made of *H* cells (or beams) with radius *R*, whereby *N* users are simultaneously active within each cell (or beam). For explanatory purposes, in the following we will refer to a cellular mobile radio network, like that depicted in Figure 2-14, with *H* =7 hexagonal shaped cells, whereby a *Base Station* (BS) is placed at the center of each cell.

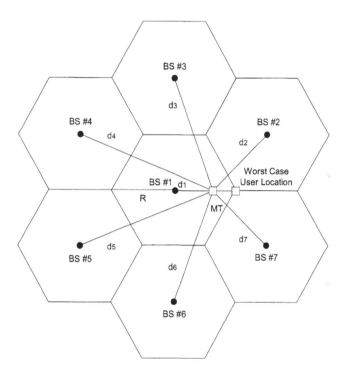

Figure 2-14. Geometry of a cellular network.

We start our analysis with the downlink. Each BS transmits a CDMA signal made of *N* traffic channels with synchronous orthogonal spreading so that the resulting multiuser traffic signal originated from each cell is similar to that in (2.70). Notice however that every BS assigns the same power to all of the signals. The universal frequency re-use causes the *Mobile Terminal* (MT) located inside cell 1 (the reference cell) to receive *H* multiplex signals in S-CDMA format arriving from all the BSs of the network. We remark that owing to different propagation times and lack of synchronization among the BSs, the overall signal received by the MT is made of an *asynchronous* combination of the *H* multiplex signal from the BSs

$$\tilde{r}(t) = \sum_{h=1}^{H}\sum_{i=1}^{N} A^{(h)} \cdot e^{-j(2\pi\Delta f_h t + \theta_h)}$$

$$\cdot \sum_{k=-\infty}^{\infty} \tilde{d}_{\{k\}_M}^{(i,h)} \cdot c_{|k|_L}^{(i,h)} \cdot g_T\left(t - kT_c - \tau_h\right) + \tilde{w}(t), \tag{2.113}$$

where a two-index notation $^{(i,h)}$ is used for each traffic signal to denote both the spreading code (index i) and the cell/beam (index h). We also denoted with $A^{(h)}$ the amplitude of the traffic channel received from the generic hth cell (see definition (2.71)), and with τ_h, θ_h, and Δf_h, the timing, carrier phase, and carrier frequency offsets, respectively, of the CDMA multiplex from cell h with respect to that of the signal received from the reference cell (cell 1). We have then $\tau_1 = 0$, $\theta_1 = 0$, and $\Delta f_1 = 0$. We can decompose (2.113) as

$$\tilde{r}(t) = A^{(1)} \cdot \sum_{k=-\infty}^{\infty} \tilde{d}_{\{k\}_M}^{(1,1)} \cdot c_{|k|_L}^{(1,1)} \cdot g_T\left(t - kT_c\right)$$

$$+ \sum_{i=2}^{N} A^{(1)} \cdot \sum_{k=-\infty}^{\infty} \tilde{d}_{\{k\}_M}^{(i,1)} \cdot c_{|k|_L}^{(i,1)} \cdot g_T\left(t - kT_c\right)$$

$$+ \sum_{h=2}^{H}\sum_{i=1}^{N} A^{(h)} \cdot e^{-j(2\pi\Delta f_h t + \theta_h)} \sum_{k=-\infty}^{\infty} \tilde{d}_{\{k\}_M}^{(i,h)} \cdot c_{|k|_L}^{(i,h)} \cdot g_T\left(t - kT_c - \tau_h\right)$$

$$+ \tilde{w}(t), \tag{2.114}$$

where the first term at the right hand side represents the useful traffic signal, the second one the *intra-cell* MAI

$$\tilde{b}^{(\text{intra})}(t) = \sum_{i=2}^{N} A^{(1)} \cdot \sum_{k=-\infty}^{\infty} \tilde{d}_{\{k\}_M}^{(i,1)} \cdot c_{|k|_L}^{(i,1)} \cdot g_T\left(t - kT_c\right), \tag{2.115}$$

and the third one the *inter-cell* MAI

$$\tilde{b}^{(\text{inter})}(t) = \sum_{h=2}^{H}\sum_{i=1}^{N} A^{(h)} \cdot e^{-j(2\pi\Delta f_h t + \theta_h)}$$

$$\cdot \sum_{k=-\infty}^{\infty} \tilde{d}_{\{k\}_M}^{(i,h)} \cdot c_{|k|_L}^{(i,h)} \cdot g_T\left(t - kT_c - \tau_h\right). \tag{2.116}$$

Applying the same kind of processing as discussed in (2.72)–(2.85) we obtain the samples at the chip matched filter output from channel 1 (see (2.74))

$$\tilde{y}_m = A^{(1)} \cdot \tilde{d}^{(1,1)}_{\{m\}_M} \cdot c^{(1,1)}_{|m|_L} + \tilde{b}^{(\text{intra})}_m + \tilde{b}^{(\text{inter})}_m + \tilde{n}_m , \tag{2.117}$$

where $\tilde{b}^{(\text{intra})}_m = \tilde{b}^{(\text{intra})}(t_m)$ and $\tilde{b}^{(\text{inter})}_m = \tilde{b}^{(\text{inter})}(t_m)$. Similarly the decision variable is (see (2.85))

$$\tilde{z}^{(1,1)}_k = A^{(1)} \cdot \tilde{d}^{(1,1)}_k + \tilde{\beta}^{(\text{intra})}_k + \tilde{\beta}^{(\text{inter})}_k + \tilde{v}_k , \tag{2.118}$$

where the terms $\tilde{\beta}^{(\text{intra})}_k$ and $\tilde{\beta}^{(\text{inter})}_k$ represent the intra- and inter-cell MAI, respectively (see (2.82)). The use of orthogonal spreading codes eliminates the effect of intra-cell MAI ($\tilde{\beta}^{(\text{intra})}_k = 0$) and the detection strobe simplifies then to

$$\tilde{z}^{(1,1)}_k = A^{(1)} \cdot \tilde{d}^{(1,1)}_k + \tilde{\beta}^{(\text{inter})}_k + \tilde{v}_k . \tag{2.119}$$

Furthermore, in the case of long pseudo-random spreading codes and large number of active users, the inter-cell MAI contribution can be modeled as a complex Gaussian random variable $\tilde{\beta}^{(\text{inter})}_k = \beta^{(\text{inter})}_{I,k} + j\beta^{(\text{inter})}_{Q,k}$ whose I/Q components are independent, identically distributed, zero mean Gaussian random variables with variance (see (2.84))

$$\sigma^2_{\beta^{(\text{inter})}_I} = \sigma^2_{\beta^{(\text{inter})}_Q} = E\left\{ \beta^{(\text{inter})2}_{I,m} \right\} = I^{(\text{inter})}_0 / T_s , \tag{2.120}$$

where we introduced the PSD of the inter-cell CDMA interference $I^{(\text{inter})}_0$. Similarly to (2.89) we obtain

$$\sigma^2_{\beta^{(\text{inter})}_I} = \sigma^2_{\beta^{(\text{inter})}_Q} = I^{(\text{inter})}_0 / T_s = P^{(\text{inter})}_{\text{MAI}} / M \tag{2.121}$$

where the power of the inter-cell MAI is given by

$$P^{(\text{inter})}_{MAI} = \sum_{h=2}^{H} \sum_{i=1}^{N} P^{(h)}_s = N \cdot \sum_{h=2}^{H} P^{(h)}_s \tag{2.122}$$

and the signal powers $P^{(h)}_s$ are related to the received signal amplitudes A_h as in (2.71). In a typical urban environment it is found that the power of a radio signal decays with the distance from the source according to the following law [Sei91]

$$P^{(h)}_s = K / d^{\zeta}_h , \tag{2.123}$$

where d_h represents the distance between the hth BS and the MT, while K is a constant factor depending on the transmitter power level, antennas gains and carrier frequency, which can be therefore assumed equal for all of the signals.

The exponent ζ is the so called *path loss exponent*, and it is found to assume values in the range $2 \div 8$, depending on the kind of propagation environment [Lee93]. A typical value for urban areas is $\zeta = 4$.

In the case of an user located at distance d_1 from the reference BS as in Figure 2-14, we find the following distances [Gia97] measured with respect to the BSs of the surrounding cells, and expressed as a function of d_1

$$d_2(d_1) = d_7(d_1) = \sqrt{3R^2 + d_1^2 - 3Rd_1},$$
$$d_3(d_1) = d_6(d_1) = \sqrt{3R^2 + d_1^2}, \tag{2.124}$$
$$d_4(d_1) = d_5(d_1) = \sqrt{3R^2 + d_1^2 + 3Rd_1},$$

and the received power levels become

$$P_s^{(2)} = P_s^{(7)} = K/d_2^\zeta,$$
$$P_s^{(3)} = P_s^{(6)} = K/d_3^\zeta, \tag{2.125}$$
$$P_s^{(4)} = P_s^{(5)} = K/d_4^\zeta,$$

The MAI power (2.122) is

$$P_{\text{MAI}}^{(\text{inter})} = 2N \cdot \left(P_s^{(2)} + P_s^{(3)} + P_s^{(4)} \right) \tag{2.126}$$

and the useful to total interfering power ratio (2.111) is

$$\left. \frac{C}{I} \right|_{tot} = \frac{P_s^{(1)}}{P_{\text{MAI}}^{(\text{inter})}} = \frac{P_s^{(1)}}{2N \cdot \left(P_s^{(2)} + P_s^{(3)} + P_s^{(4)} \right)}. \tag{2.127}$$

The BER in the presence of inter-cell MAI is obtained from (2.88), by simply substituting I_0 with $I_0^{(\text{inter})}$ as in (2.121). Combining (2.122) with (2.125) we obtain the following BER

$$P(e) = Q\left(\sqrt{\frac{2E_b}{N_0}} \cdot \frac{1}{\sqrt{1 + \dfrac{2N \cdot \sum\limits_{h=2}^{4}\left(d_1/d_h(d_1)\right)^{\varsigma}}{M} \dfrac{2E_b}{N_0}}}\right), \tag{2.128}$$

which depends on the distance d_1 of the MT with respect to the reference BS. We evaluate first the error probability (2.128) for a MT located in the *Worst Case* (WC) user location, i.e., in the farthest point from the reference BS (see Figure 2-14). A user placed in such a location receives the minimum power of the useful signal from the reference BS (no. 1) and the maximum of interference from the closest interfering BSs (no. 2 and no. 7). Letting $d_1 = R$, with some geometry, the 'BS to MT' distances d_h (2.124) are found to be

$$d_2 = d_7 = R,$$
$$d_3 = d_6 = 2R, \tag{2.129}$$
$$d_4 = d_5 = \sqrt{7}R.$$

If we are interested in a less pessimistic case, or in a sort of *Average Case* (AC), we need a statistical model for the spatial distribution of the MTs within the reference cell. A reasonable assumption consists in considering all the locations within the cell as equally probable. Toggling now, for the sake of simplicity, to a circular cell model such as that in Figure 2-15, the probability that the MT lies within any region of area S all within the cell is given by

$$P(S) = \frac{S}{\pi R^2}. \tag{2.130}$$

Also, the probability that the MT is located at a distance x ($0 \le x \le R$) from the BS is given by the probability that it lies inside the circular corona having infinitesimal width dx and radius x, represented by the grey shaded region in Figure 2-15. Such probability is given by

$$dP = \frac{dS}{\pi R^2} = \frac{2\pi x\, dx}{\pi R^2} = \frac{2x}{R^2}dx. \tag{2.131}$$

The probability density function of the random variable X representing the distance from the BS to the MT is then

$$p_X(x) = \frac{dP}{dx} = \frac{2x}{R^2}, \quad 0 \le x \le R.$$ (2.132)

Finally, we compute the mean value of the distance

$$\eta_X = E\{X\} = \int_{-\infty}^{\infty} x \, p_X(x) \, dx = \int_0^R \frac{2x^2}{R^2} \, dx = \frac{2}{3} R.$$ (2.133)

Letting $d_1 = \eta_X = 2R/3$, we can now evaluate the BER (2.128) for the AC user location. With some geometry the 'BS to MT' distances d_h (2.124) are now found to be

$$d_2 = d_7 = \sqrt{5}R/2,$$
$$d_3 = d_6 = \sqrt{31}R/2,$$ (2.134)
$$d_4 = d_5 = 7R/2.$$

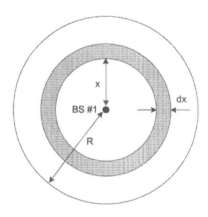

Figure 2-15. Circular cell model.

The uplink is typically based upon asynchronous random access from MTs to the BSs, and therefore any receiving BS experiences both intra- and inter-cell asynchronous MAI. The overall signal received by the reference BS in the uplink is then made of an *asynchronous* combination of the signals originated from all of the MTs active users within all the cells and can be put in a form similar to (2.106)

$$\tilde{r}(t) = \sum_{h=1}^{H}\sum_{i=1}^{N} A^{(i,h)} \cdot e^{-j\left(2\pi\Delta f_{i,h}t + \theta_{i,h}\right)}$$

$$\cdot \sum_{k=-\infty}^{\infty} \tilde{d}_{\{k\}_M}^{(i,h)} \cdot c_{|k|_L}^{(i,h)} \cdot g_T\left(t - kT_c - \tau_{i,h}\right) + \tilde{w}(t). \tag{2.135}$$

Notice that owing to the lack of synchronicity amongst all the transmitters and to the different user positions, any MT contribution is characterized by a different amplitude $A^{(i,h)}$, timing $\tau_{i,h}$, carrier phase $\theta_{i,h}$, and carrier frequency $\Delta f_{i,h}$ offsets. We can decompose (2.135) as

$$\tilde{r}(t) = A^{(1,1)} \cdot \sum_{k=-\infty}^{\infty} \tilde{d}_{\{k\}_M}^{(1,1)} \cdot c_{|k|_L}^{(1,1)} \cdot g_T\left(t - kT_c\right) +$$

$$+ \sum_{i=2}^{N} A^{(i,1)} \cdot e^{-j\left(2\pi\Delta f_{i,1}t + \theta_{i,1}\right)} \sum_{k=-\infty}^{\infty} \tilde{d}_{\{k\}_M}^{(i,1)} \cdot c_{|k|_L}^{(i,1)} \cdot g_T\left(t - kT_c - \tau_{i,1}\right)$$

$$+ \sum_{h=2}^{H}\sum_{i=1}^{N} A^{(i,h)} \cdot e^{-j\left(2\pi\Delta f_{i,h}t + \theta_{i,h}\right)} \sum_{k=-\infty}^{\infty} \tilde{d}_{\{k\}_M}^{(i,h)} \cdot c_{|k|_L}^{(i,h)} \cdot g_T\left(t - kT_c - \tau_{i,h}\right)$$

$$+ \tilde{w}(t), \tag{2.136}$$

where the first term in the right hand side represents the useful traffic signal, the second one the *intra-cell* MAI

$$\tilde{b}^{(\text{intra})}(t) = \sum_{i=2}^{N} A^{(i,1)} \cdot e^{-j\left(2\pi\Delta f_{i,1}t + \theta_{i,1}\right)} \sum_{k=-\infty}^{\infty} \tilde{d}_{\{k\}_M}^{(i,1)} \cdot c_{|k|_L}^{(i,1)} \cdot g_T\left(t - kT_c - \tau_{i,1}\right) \tag{2.137}$$

and the third one the *inter-cell* MAI

$$\tilde{b}^{(\text{inter})}(t) = \sum_{h=2}^{H}\sum_{i=1}^{N} A^{(i,h)} \cdot e^{-j\left(2\pi\Delta f_{i,h}t + \theta_{i,h}\right)}$$

$$\cdot \sum_{k=-\infty}^{\infty} \tilde{d}_{\{k\}_M}^{(i,h)} \cdot c_{|k|_L}^{(i,h)} \cdot g_T\left(t - kT_c - \tau_{i,h}\right). \tag{2.138}$$

As is shown in Section 5 below, the MAI terms can at times overwhelm the useful signal component. To prevent this, modern CDMA systems implement some form of power control, so that all the signals originated from the MTs located within the generic hth cell are received with same amplitude, say $A^{(h)}$, by the relevant BS. The intra-cell contribution then becomes

$$\tilde{b}^{(\text{intra})}(t) = \sum_{i=2}^{N} A^{(1)} \cdot e^{-j\left(2\pi\Delta f_{i,1}t + \theta_{i,1}\right)} \sum_{k=-\infty}^{\infty} \tilde{d}_{\{k\}_M}^{(i,1)} \cdot c_{|k|_L}^{(i,1)} \cdot g_T\left(t - kT_c - \tau_{i,1}\right). \tag{2.139}$$

By applying a procedure similar to that described by (2.72)–(2.85), and referring to the useful traffic channel, represented by user 1 of cell 1, we can derive the following expression for the decision strobe (which is formally identical to (2.118))

$$\tilde{z}_k^{(1,1)} = A^{(1)} \cdot \tilde{d}_k^{(1,1)} + \tilde{\beta}_k^{(\text{intra})} + \tilde{\beta}_k^{(\text{inter})} + \tilde{v}_k, \tag{2.140}$$

where the samples $\tilde{\beta}_k^{(\text{intra})}$ and $\tilde{\beta}_k^{(\text{inter})}$ represent again the intra- and inter-cell MAI residual disturbance, respectively. Owing to the asynchronous random access from MTs to the BSs adopted in the uplink, intra-cell orthogonality can no longer be invoked, and, differently from (2.119), $\tilde{\beta}_k^{(\text{intra})}$ is not null.

Let us start by considering for now the issue of an uplink affected by intra-cell interference only, as in the case of a single-cell scenario. In the usual case of long codes and large number of active users, the inta-cell contribution can be modeled as a complex Gaussian random variable denoted as $\tilde{\beta}_k^{(\text{intra})} = \beta_{I,k}^{(\text{intra})} + j\beta_{Q,k}^{(\text{intra})}$ whose I/Q components are independent identically distributed zero mean Gaussian random variables with variance (see (2.84))

$$\sigma_{\beta_I^{(\text{intra})}}^2 = \sigma_{\beta_Q^{(\text{intra})}}^2 = E\left\{ \beta_{I,m}^{(\text{intra})2} \right\} = \frac{I_0^{(\text{intra})}}{T_s}, \tag{2.141}$$

where $I_0^{(\text{intra})}$ is the PSD of the intra-cell CDMA interference. Similarly to (2.89) we obtain

$$\sigma_{\beta_I^{(\text{intra})}}^2 = \sigma_{\beta_Q^{(\text{intra})}}^2 = \frac{I_0^{(\text{intra})}}{T_s} = \frac{P_{\text{MAI}}^{(\text{intra})}}{M}, \tag{2.142}$$

where $P_{\text{MAI}}^{(\text{intra})}$ is the power of the intra-cell MAI, which, in the case of perfect power control, is given by

$$P_{\text{MAI}}^{(\text{intra})} = \sum_{i=2}^{N} P_s^{(1)} = (N-1) P_s^{(1)}, \tag{2.143}$$

and the signal power $P_s^{(1)}$ is related to the received signal amplitude A_1 as in (2.71). The useful to intra-cell interfering power ratio (2.111) is now

$$\left. \frac{C}{I} \right|_{\text{intra}} = \frac{P_s^{(1)}}{P_{\text{MAI}}^{(\text{intra})}} = \frac{1}{N-1}. \tag{2.144}$$

The BER for the uplink of a single-cell case, i.e., in the presence of intra-cell MAI only, is obtained from (2.88) by simply substituting I_0 with $I_0^{(\text{intra})}$ as in (2.142–143)

$$P(e) = Q\left(\sqrt{\frac{2E_b}{N_0}} \cdot \frac{1}{\sqrt{1 + \dfrac{(N-1)}{M}\dfrac{2E_b}{N_0}}} \right) \qquad (2.145)$$

Let us consider now the extension of the analysis above to the more general case of an uplink in a multi-cell scenario. It must be remarked that power control makes the interfering power received at the reference BS from the surrounding cells to be dependent on the random distance between the interfering MT and the BS serving the cell containing that particular MT. The issue of MAI evaluation in the uplink of a multi-cell network gets rather involved and will not be presented here [New94]. We will just remark that the outcome of such investigation is the evaluation of a coefficient, the *inter-cell interference factor*, defined as

$$\varphi = \frac{P_{\text{MAI}}^{(\text{inter})}}{P_{\text{MAI}}^{(\text{intra})}} \qquad (2.146)$$

so that the total MAI experienced by the reference BS can be expressed as

$$P_{\text{MAI}} = P_{\text{MAI}}^{(\text{intra})} + P_{\text{MAI}}^{(\text{inter})} = (1 + \varphi) P_{\text{MAI}}^{(\text{intra})}. \qquad (2.147)$$

The value of φ depends on the path loss exponent ζ as shown in Figure 2-16. For $\zeta = 4$ and in the presence of perfect power control, we have $\varphi \cong 0.55$ [New94], [Vit95]. From (2.143) we obtain

$$P_{\text{MAI}} = (1 + \varphi)(N - 1) P_s^{(1)}, \qquad (2.148)$$

and the BER is obtained by a simple modification of (2.145)

$$P(e) = Q\left(\sqrt{\frac{2E_b}{N_0}} \cdot \frac{1}{\sqrt{1 + \dfrac{(1+\varphi)\cdot(N-1)}{M}\dfrac{2E_b}{N_0}}} \right) \qquad (2.149)$$

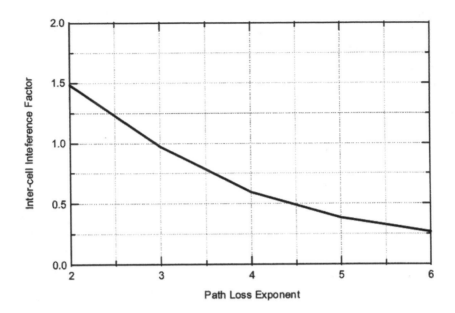

Figure 2-16. Inter-cell interference factor vs. path loss exponent

Finally, Figure 2-17 presents the BER performance relevant to some downlink and uplink configurations, evaluated for $L = 64$, $E_b / N_0 = 9.6$ dB and $\zeta = 4$. Crosses and triangles refer to the downlink for the worst and average case, respectively, and have been plotted by using (2.128) with the settings (2.129) for the WC and (2.134) for the AC. Dots and squares refer to the uplink, under the hypothesis of perfect power control, for the single and multi-cell scenario respectively and were derived using (2.145) and (2.149) with $\varphi = 0.55$.

5.   INTERFERENCE MITIGATION RECEIVERS FOR THE DOWNLINK

The discussion in the previous Section has outlined the main issue that affects CDMA systems as far as capacity and/or quality of service are concerned, namely, interference. This is certainly true for the asynchronous uplink wherein interference in the form of MAI is generated within one's own cell (intra-cell interference), but applies to the downlink as well, where asynchronous, non-orthogonal inter-cell interference is generated by adjacent cells operating on same carrier frequency in the same coverage area.

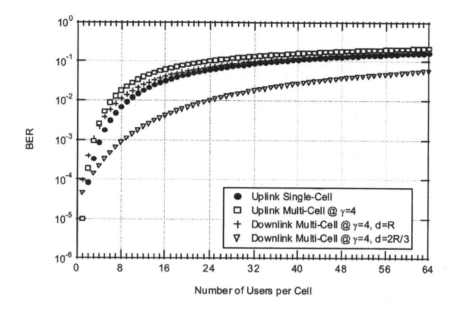

Figure 2-17. BER vs. cell loading for downlink and uplink configurations

The seminal work by Verdù [Ver86] and almost two decades of related research have shown how to cope with such issues [Due95], [Mos96], [Las97], [Ver97], [Hon98]. MAI in the uplink can be counteracted by the adoption at the BS of a suited *joint* or *MultiUser Data Detection* scheme (MUD) that jointly performs demodulation of *all* of the uplink data streams in a single, centralized signal processing unit. With centralized detection the presence of a certain amount of MAI coming from interfering channels can be accounted for when demodulating the useful, intended channel. This standpoint applies reciprocally to *all* of the channels, leading to the above-mentioned notion of 'multiuser' demodulation. To be more specific, a conventional arrangement of the BS channel demodulator is shown in Fig 2-18. The data demodulation unit is just architected as an array of independent single-channel demodulators in the form of conventional matched filter detectors, like that depicted in Figure 2-10. Those detectors are optimum in the AWGN environment – they simply ignore the issue of MAI and lead to strong sub-optimality.

In contrast, a multiuser detector is a centralized data demodulation unit whose general scheme is depicted in Figure 2-19. By concurrently observing all of the matched filter outputs, MAI can be taken into account when detecting data of each channel, and it can be mitigated or 'cancelled' by suitable signal processing. This 'interference cancellation' feature of MUD relaxes the demand on the cross-correlation properties of the user signature sequences since good BER performance can be attained even in the presence of non-orthogonal, asynchronous MAI. This gives a new insight to the issue of CDMA detection, in that it suggests a way of increasing the capacity of the CDMA system while keeping the quality of the link (i.e., its BER) constant. This two-fold leap forward in the performance of the CDMA receiver, not surprisingly, comes at the expense of a substantial increase in the complexity (signal processing power) of the demodulator. The front end of the centralized MUD scheme in Figure 2-19 consists of the same bank of N correlators with the individual N user signature waveforms that we find in the multiple conventional CRs in Figure 2-18. The real core of MUD lies in the elaborate post-processing of the array of matched filter outputs (sufficient statistics) [Mos96].

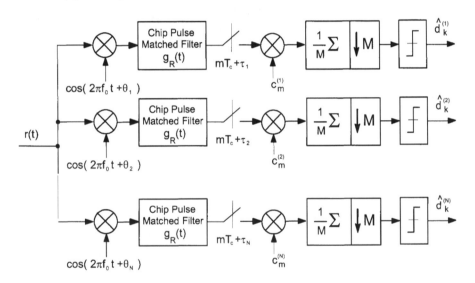

Figure 2-18. Bank of conventional single-user detectors in the uplink

The optimum AWGN MUD scheme originally proposed and analyzed by Verdù [Ver86] encompasses a *Viterbi Algorithm* (VA) to perform parallel maximum likelihood sequence estimation. Since the number of states in the trellis of the VA is equal to 2^N, a receiver with this optimum structure has a complexity that is exponential in the number of users N, and therefore it does not easily lend itself to practical implementations. Several different suboptimum MUD structures have therefore appeared since then, all of them attempting to reduce the estimation complexity by replacing the Viterbi decoder by a different device (decorrelating detector, successive cancellations, and so on [Mos96]).

A further constraint which has to be satisfied to allow implementation of MUD is that the CDMA signals to be used in the multiple access network bear *short codes*, i.e., $L = M$, so the cross-correlations between the different spreading codes are well defined and possibly different from each other. MUD in its different forms relies in fact on the *cyclostationarity* on a code period of CDMA signals, that cannot be exploited in the case of long spreading codes, as is unfortunately the case of the IS-95 uplink. The long code is introduced in IS-95 to *randomize* the user signals and to make MAI look as much white and Gaussian as possible (with complete destruction of cyclostationarity of MAI). MAI is then dealt with by means of powerful low rate error correcting codes that tame out the influence of interference. This is a radically different 'CDMA philosophy' [Vem96] with respect to short codes and MUD.

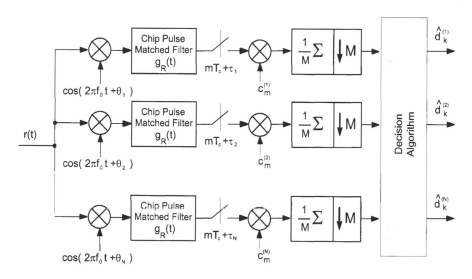

Figure 2-19. General scheme of a multiuser detector.

To give a hint of how the MUD receiver work, let us assume that we intend to demodulate a three-channel CDMA signal with real spreading of real-valued symbols. To simplify matters we will also assume that the three channels (codes) are *synchronous*, but they are *not* orthogonal. Although this is not fully representative of the situation encountered in the uplink (asynchronous non-orthogonal), it leads to similar results. Assuming ideal code timing synchronization, the symbol rate sampled outputs of the code matched filters in the front end of the MUD in Figure 2-4 are as follows

$$z_k^{(1)} = A^{(1)}d_k^{(1)} + \rho_{1,2}A^{(2)}d_k^{(2)} + \rho_{1,3}A^{(3)}d_k^{(3)} + v_k^{(1)},$$
$$z_k^{(2)} = \rho_{2,1}A^{(1)}d_k^{(1)} + A^{(2)}d_k^{(2)} + \rho_{2,3}A^{(3)}d_k^{(3)} + v_k^{(2)}, \qquad (2.150)$$
$$z_k^{(3)} = \rho_{3,1}A^{(1)}d_k^{(1)} + \rho_{3,2}A^{(2)}d_k^{(2)} + A^{(3)}d_k^{(3)} + v_k^{(3)},$$

where $\rho_{m,k} = \rho_{k,m}$ is the shorthand notation for *cross-correlation* coefficient $R_{c^{(m)}c^{(k)}}$ between the two spreading codes of channel m and channel k, $A^{(i)}$ is the amplitude of signal on channel i, $d_k^{(i)}$ is the kth data symbol on channel i, and $v_k^{(i)}$ is the ith noise component. The effect of MAI is apparent, and it is also clear that it can be potentially destructive. Assume for instance that $\rho_{1,2} \neq 0$ (codes 1 and 2 are not orthogonal) and that $A^{(2)} \gg A^{(1)}$: the MAI term $\rho_{1,2}A^{(2)}d_k^{(2)}$ in (2.150) may overwhelm the useful term $A^{(1)}d_k^{(1)}$ for data detection of channel 1. This phenomenon is called the *near–far effect*: user 2 can be considered as located near the receiver in the BS, thus received with a large amplitude, whilst user 1 (the one we intend to demodulate) is the far user and is weaker than user 2. Generalizing to N users, equations (2.150) can be easily cast into a simple matrix form. If we arrange the cross-correlation coefficients $\rho_{i,k}$ into the correlation matrix

$$
\mathbf{R} = \begin{pmatrix} 1 & \rho_{1,2} & \cdots & \rho_{1,N} \\ \rho_{2,1} & 1 & & \rho_{2,N} \\ \vdots & & \ddots & \vdots \\ \rho_{N,1} & \rho_{N,2} & \cdots & 1 \end{pmatrix} \tag{2.151}
$$

and if we also introduce the diagonal matrix of the user amplitudes $\mathbf{A} = \mathrm{diag}\left\{ A^{(1)}, A^{(2)}, ..., A^{(N)} \right\}$, we have

$$
\mathbf{z} = \mathbf{A}\mathbf{R}\mathbf{d} + \mathbf{v} \tag{2.152}
$$

where the column vectors $\mathbf{z}$, $\mathbf{d}$, and $\mathbf{v}$ simply collect the respective samples of received signal, data, and noise. A simple multiuser detector is the *decorrelating detector* that applies a linear transformation to vector $\mathbf{z}$ to provide N 'soft' decision variables relevant to the N data bits to be estimated. Collecting such N decision variables $g_k^{(i)}$ into vector $\mathbf{g}$, the linear *joint* transformation on the matched filter output vector $\mathbf{z}$ is just

$$
\mathbf{g} = \mathbf{R}^{-1}\mathbf{z} \tag{2.153}
$$

where $\mathbf{R}^{-1}$ is the *decorrelating matrix*, so that

$$
\mathbf{g} = \mathbf{R}^{-1}\left(\mathbf{A}\mathbf{R}\mathbf{d} + \mathbf{v}\right) = \mathbf{A}\mathbf{d} + \mathbf{R}^{-1}\mathbf{v} =
$$
$$
= \mathrm{diag}\left\{ A_1 d_k^{(1)}, A_2 d_k^{(2)}, ..., A_N d_k^{(N)} \right\} + \mathbf{v}' \tag{2.154}
$$

We have thus

$$\mathbf{g} = \left[g_k^{(1)}, g_k^{(2)}, ..., g_k^{(N)} \right]^T \tag{2.155}$$

where the superscript ^{T} denotes transposition, $g_k^{(i)} = A^{(i)} d_k^{(i)} + v_k'^{(i)}$, and $v_k'^{(i)}$ is just a noise component. It is apparent that the MAI has been *completely cancelled* (provided that the correlation matrix is invertible) or, in other words, the different channels have been *decorrelated*. The drawback is an effect of *noise enhancement* owed to the application of the decorrelating matrix $\mathbf{R}^{-1}$: the variance of the noise components in $\mathbf{v}'$ is in general larger than that the components in $\mathbf{v}$. Therefore, the decorrelating detector works fine only when the MAI is largely dominant over noise.

A different approach is pursued in the design of the *Minimum Mean Square Error* (MMSE) multiuser detector: the linear transformation is now with a generic $N \times N$ matrix $\mathbf{Z}$ whose components are such that the MMSE between the soft output decision variables in $\mathbf{g}$ and the vector of the data symbols is minimized

$$\mathbf{g} = \mathbf{Z}\mathbf{z} , \text{ with } \mathbf{Z} \text{ such that } E\left\{ |\mathbf{g} - \mathbf{d}|^2 \right\} = \min , \tag{2.156}$$

where $E\{\cdot\}$ denotes statistical expectation. Solving for $\mathbf{Z}$ we have

$$\mathbf{Z} = \left(\mathbf{R} + \sigma_v^2 \mathbf{A}^{-2} \right)^{-1} , \tag{2.157}$$

with σ_v^2 indicating the variance of the noise components in (2.150). The MMSE detector tries to optimize the linear transformation both with respect to MAI and to noise. If noise is negligible with respect to MAI, matrix (2.157) collapses into the decorrelating matrix $\mathbf{R}^{-1}$. Vice versa, if the MAI is negligible, the matrix $\mathbf{Z}$ is diagonal and collapses just into a set of scaling factors on the matched filters output that do not affect data decisions at all (and in fact in the absence of MAI, the outputs of the matched filters are the optimum decision variables without any need of further processing).

From this short discussion about MUD it is clear that in general such techniques are quite challenging to implement, either because they require non-negligible processing power (for instance, to invert the decorrelating or the MMSE matrices), and because they also call for *a priori* knowledge or real time estimation of signal parameters, such as the correlation matrix. But the potential performance gain of MUD had also an impact on the standardization of 3G systems (UMTS in Europe), in that an option for short codes in the downlink was introduced just to allow for the application of such techniques in the BS [Ada98], [Dah98], [Oja98], [Pra98].

How can MUD or related techniques be applied to the *downlink* of the wireless system? Multiuser detection in the user terminal (the mobile phone) has no meaning at all, since the UT is by definition a single-channel demodulator. Also, if channel equalization is good, hence channel distortions are negligible, no MAI is experienced in the downlink, since the channelization codes are orthogonal. Nonetheless, the downlink experiences inter-cell interference, especially when the UT is close to a neighboring cell boundary. Therefore a single-channel *Interference Mitigating Detector* (IMD) is something the downlink would surely benefit from. Our previous consideration about the two-sided effect of interference mitigation applies to the downlink as well: the IMD can be used either to improve the quality of the link for a given level of interference, or it can be used as an instrument to increase network capacity for a given quality of the link.

How can we implement an IMD? We start by recalling the signal samples at the chip matched filter output (see (2.117))

$$\tilde{y}_m = A^{(1)} \cdot \tilde{d}^{(1,1)}_{\{m\}_L} \cdot c^{(1,1)}_{|m|_L} + \tilde{b}^{(\text{intra})}_m + \tilde{b}^{(\text{inter})}_m + \tilde{n}_m \tag{2.158}$$

where $\tilde{b}^{(\text{intra})}_m$ and $\tilde{b}^{(\text{intra})}_m$ denote the mth sample of the intra- and inter-cell interference term, respectively. The two terms together make up a disturbance term that is *independent* of the useful signal component and adds up to the background noise. Also, the total disturbance term $D_m = \tilde{b}^{(\text{intra})}_m + \tilde{b}^{(\text{inter})}_m$ is *not* white as, in contrast, $\tilde{n}_m$ is: the standard CR with the code matched filter (or the despreader–accumulator cascade) is no longer optimum. It makes sense therefore correlating the received signal samples with a set of coefficients that is *not* equal to the values of the spreading code $c^{(1)}_m$. The decision variable for the kth data symbol on channel 1 will be thus equal to

$$\tilde{z}^{(1)}_k = \frac{1}{L} \cdot \sum_{m=0}^{L-1} \tilde{y}_{kL+m} \cdot h^{(1)}_m \tag{2.159}$$

where the coefficients h_m (from now on we will omit the superscript $^{(1)}$ for simplicity) have to be designed according to a suited optimization rule. Equation (2.159) can also be interpreted as the response to the input $\tilde{y}_m$ of a linear filter whose coefficients are just h_m, downsampled to the symbol rate. The simplest yet most effective criterion to design the filter coefficients is again the MMSE rule, of course this time in a simplified single-channel version with respect to (2.156)

$$\tilde{z}_k = \frac{1}{L} \cdot \sum_{m=0}^{L-1} \tilde{y}_{kL+m} \cdot h_m \quad ,$$

with $h_0,...,h_{L-1}$ such that $E\left\{\left|\tilde{z}_k - \tilde{d}_k^{(1)}\right|^2\right\} = \min$ (2.160)

We have to face an issue similar to that encountered in the MMSE MUD; specifically, how to set the filter coefficients in order to solve the minimization problem just stated. The solution of this issue leads to the concept of an *adaptive* detector, whose coefficients are adapted in real time so as to minimize the mean square error and thus build 'on the fly' an optimum linear detector for the configuration of interference that the reference channel is experiencing. We skip the detailed solution of the minimization problem [Mad94] to report here the *recursive* equation that, starting from arbitrary values of the filter coefficients, allows the synthesis of the optimum detector configuration

$$\mathbf{h}(k+1) = \mathbf{h}(k) - \gamma\left(\tilde{z}_k - \tilde{d}_k^{(1)}\right)\tilde{\mathbf{y}}(k)^*,$$ (2.161)

where the L-dimensional vectors $\mathbf{h}(k)$ and $\tilde{\mathbf{y}}(k)$ group the filter coefficients at time k and the received signal samples $\tilde{y}_{kL+m}$, $m = 0,1,...,L-1$, respectively. In (2.161) γ is the recursion step size, to be set a compromise between fast acquisition (large γ) and small steady state fluctuations (small γ). The drawback of the adaptive MMSE detector (2.161) is that recursive adaptation of the coefficient vector $\mathbf{h}(k)$ calls for the knowledge of the transmitted data $\tilde{d}_k^{(1)}$ to compute the error $e_k = \tilde{z}_k - \tilde{d}_k^{(1)}$. This *Data Aided* (DA) approach can be adopted if a set of pilot symbols is organized into a preamble known to the receiver in an initial *training phase*. At the end of the training phase, the detector coefficients are 'frozen' and true data detection starts. The detector operates thus in *Decision Directed* (DD) mode. If MAI is time varying, adaptation of the coefficients must be periodically carried out each time a new preamble of known data appears in the signal framing. Of course, this has an impact on the efficiency of the communication link, since the preamble data does not convey any information, and contribute to the overall data framing overhead. In addition, the recursion (2.161) may require a large number of training symbols to attain a steady state condition (long acquisition time), making the adoption of a data aided approach impractical.

Therefore it makes sense to revert to a *blind* approach that does not require the insertion of any pilot symbols or preambles in the data stream. This privileges framing efficiency and is also robust in terms of acquisition/reacquisition capability. The criterion to be adopted to satisfy this requirement is the *minimization of the Mean Output Energy (Minimum MOE,*

MMOE) of the detector instead of the minimization of the squared error as before

$$\tilde{z}_k = \frac{1}{L} \cdot \sum_{m=0}^{L-1} \tilde{y}_{kL+m} \cdot h_m \,, \quad h_0,...,h_{L-1} \text{ such that } E\left\{\left|\tilde{z}_k\right|^2\right\} = \min. \qquad (2.162)$$

The rationale behind this criterion is that by minimizing the output energy the influence of MAI is minimized as well. Of course, we have to add some additional constraints to this minimization problem, otherwise the solution is a trivial one: all coefficients are equal to 0. The trick to avoid the coefficients array $\mathbf{h}(k)$ collapsing to $\mathbf{0}$ is the *anchoring* of its value to the value it would have in the absence of interference. We know that with no MAI the optimum receiver is the conventional correlator, so that in those conditions $\mathbf{h}(k) = \mathbf{c}$, where $\mathbf{c}$ is the L-dimensional array containing the code chips c_i ($i = 0,...L-1$) of the desired user. In the general case we set

$$\mathbf{h}(k) = \mathbf{c} + \mathbf{x}(k) \qquad (2.163)$$

where the constraint is that $\mathbf{c}$ and $\mathbf{x}(k)$ be *orthogonal*: $\mathbf{c}^T \mathbf{x}(k) = 0$ (the superscript ^{T} denotes matrix transposition). This is what we called the 'anchoring', and this is also what prevents the coefficients from converging towards $\mathbf{0}$. This simple idea, which was introduced by Honig, Madhow, and Verdù [Hon95], led to the development of what is called the *Extended, Complex-valued, Blind, Anchored, Interference mitigating Detector* (EC-BAID). Design of the detector (and adaptivity of the detector as well) is now transferred to design and adaptivity of the code orthogonal vector $\mathbf{x}$.

In a sense, decomposition (2.163) can lead us to interpret the MMOE detector as the superposition of *two* detectors: the one characterized by the set of coefficients $\mathbf{c}$ is the conventional detector which is optimum for the AWGN channel. The other 'additional' detector $\mathbf{x}$ gives the additional feature of interference mitigation. It can be shown [Hon95] that the MMOE solution for $\mathbf{x}$ gives also the MMSE solution for $\mathbf{h}$, i.e., $\mathbf{h}_{\text{MMSE}} = \mathbf{c} + \mathbf{x}_{\text{MOE}}$. The resulting recursive equation for the vector $\mathbf{x}$ is

$$\mathbf{x}(k+1) = \mathbf{x}(k) - \gamma\, \tilde{z}(k) \left[\mathbf{y}(k) - \frac{\mathbf{y}(k)^T \mathbf{c}}{L} \mathbf{c} \right]^*. \qquad (2.164)$$

The second term between brackets is the orthogonal projection of the vector of received samples onto the spreading code $\mathbf{c}$. So the EC-BAID is a modified MMOE linear detector operating on the received signal, sampled at the chip rate, $\tilde{y}_m$ to yield the symbol rate signal $\tilde{z}_k$ as follows

$$\tilde{z}_k = \frac{1}{L}\mathbf{h}^e(k)^T \mathbf{y}^e(k),$$
(2.165)

where $\mathbf{y}^e(k)$ is the *extended* 3L-dimensional array of the received signal samples coming from *three* symbol periods (i.e., the current period, the leading period, and the trailing period) whose elements are denoted as y_i^e

$$\mathbf{y}^e(k) = \begin{bmatrix} \mathbf{y}_{-1}(k) \\ \mathbf{y}_0(k) \\ \mathbf{y}_1(k) \end{bmatrix} \quad \text{and} \quad \mathbf{y}_i(k) = \left[\tilde{y}_{(k+i)L}, \tilde{y}_{(k+i)L+1}, \dots, \tilde{y}_{(k+i)L+L-1} \right]^T,$$
(2.166)

and $\mathbf{h}^e(k)$ is a similarly extended array of detector coefficients. It is apparent that *extension* refers to lengthening of the observation window of the signal. Such extension is beneficial in terms of the interference rejection capability of the detector, especially for asynchronous MAI. The extended detector can be effectively implemented for real time operation according to the three-fold parallel architecture sketched in Figure 2-20, wherein the first unit processes the $(k-1)$th, the kth and the $(k+1)$th symbol periods for the detection of the kth symbol, the second unit processes the kth, the $(k+1)$th and the $(k+2)$th periods, for the detection of the $(k+1)$th symbol, and the third unit processes the $(k+1)$th, the $(k+2)$th and the $(k+3)$th periods, for the detection of the $(k+2)$th symbol. Each detector unit has the structure outlined in Figure 2-21. The final soft output data stream is obtained by sequentially selecting one of the three detector outputs at the symbol rate $1/T_s$ by means of a multiplexer. To better explain operations of the EC-BAID circuit in Figure 2-20 it is expedient to introduce a further clock reference ticking at what we call the *Super-Symbol* (SS) rate $R_{ss} = 1/(3T_s)$, i.e., once every three symbols. R_{ss} is basically the operating rate of each of the three detectors. The output of the nth detector unit ($n = 1, 2, 3$) is computed as

$$\tilde{z}(3s+n-1) = \frac{1}{L}\mathbf{h}^{e,n}(s)^T \mathbf{y}^e(3s+n-1),$$
(2.167)

with s running at super-symbol rate. To achieve blind adaptation, the complex detector coefficients are anchored to the user signature sequence as outlined above.

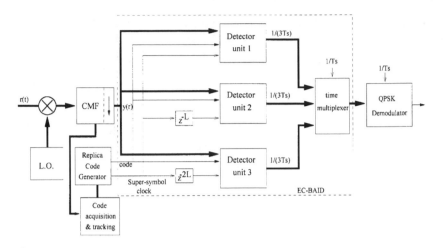

Figure 2-20. EC-BAID General Architecture.

Specifically, the extended detectors are characterized by the set of coefficients

$$
\mathbf{h}^{e,n}(s) = \mathbf{c}^e + \mathbf{x}^{e,n}(s), \quad \mathbf{c}^e = \begin{bmatrix} \mathbf{0} \\ \mathbf{c} \\ \mathbf{0} \end{bmatrix}, \quad \mathbf{x}^{e,n}(s) = \begin{bmatrix} \mathbf{x}^n_{-1}(s) \\ \mathbf{x}^n_0(s) \\ \mathbf{x}^n_1(s) \end{bmatrix}, \tag{2.168}
$$

with the 'anchor' constraint $\mathbf{c}^T \mathbf{x}^n_i = 0$, $i = -1, 0, 1$, $n = 1, 2, 3$. By trivially generalizing the recursive equation (2.164) we obtain the updating rule for the interference-mitigating vectors of the three detectors

$$
\mathbf{x}^{e,n}(s+1) = \mathbf{x}^{e,n}(s) - \gamma \mathbf{e}^{e,n}(s), \tag{2.169}
$$

with s ticking at the super-symbol rate, and where

$$
\mathbf{e}^{e,n}(s) = \begin{bmatrix} \mathbf{e}^n_{-1}(s) \\ \mathbf{e}^n_0(s) \\ \mathbf{e}^n_1(s) \end{bmatrix},
$$

$$
\mathbf{e}^n_i(s) = \tilde{z}(s) \left[\mathbf{y}^*_i(3s+n-1) - \frac{\mathbf{y}^*_i(3s+n-1)^T \mathbf{c}}{L} \mathbf{c} \right], \quad i = -1, 0, 1, \tag{2.170}
$$

γ is the adaptation step and the asterisk denotes complex conjugation. Equation (2.169) implicitly assumes that the three detector units are running independently. More architectures can be devised wherein the error control signal for the update of vectors $\mathbf{x}^{e,n}$, whose elements are denoted as x_i^e, is unique and is obtained as a combination of the partial errors $\mathbf{e}^{e,n}$ [Rom00].

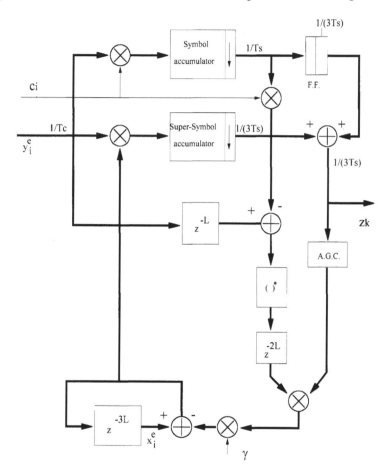

Figure 2-21. Internal structure of the three detectors in Figure 2-20.

An example of the interference mitigation capability of the EC-BAID is given in Figure 2-22. We show in the chart the BER of a CDMA receiver with asynchronous interference as a function of the number of concurrently active users. The spreading factor is 64, the spreading codes are Walsh–Hadamard with an Extended PN superimposed as scrambling code, and the users delay are uniformly spaced over one symbol interval. The curve labeled BAID is obtained with a MMOE detector observing a single-symbol period, whilst the one labeled EC-BAID is obtained with the three-symbol extended detector above. The superiority with respect to the conventional correlation receiver is apparent, although it is also apparent that when the number of channels gets close to the spreading factor even the IM detectors cannot counteract MAI any longer.

The curves in Figure 2-22 also help to explain how the IM detectors can be seen as a technological factor for increasing the network capacity in terms of number of served users per cell. Assume that we place a QoS constraint in terms of BER of the link, 10^{-2} just to be specific. The curve of the correlation receiver in Figure 2-22 says that the maximum number of users in an hypothetical cell with that spreading factor is restricted to about 7 (that is, the value on the abscissas corresponding to the specified BER). The corresponding figure on the EC-BAID curve at the same QoS is roughly 38, with more than a 5-fold capacity increase!

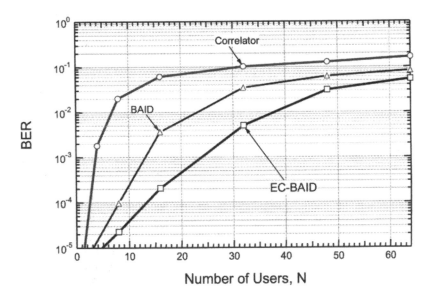

Figure 2-22. Interference-mitigating capability of the EC-BAID.

6.   A SAMPLE CDMA COMMUNICATION SYSTEM: SPECIFICATIONS OF THE MUSIC TESTBED

As an example of a practical, we present hereafter the specifications of the CDMA system envisaged in the framework of the MUSIC project, sponsored by the ESA [MUS01]. Table 2-2 contains the specifications of the CDMA signal generator, including signal format, programmability features and physical characteristics of analog modulator. Concerning signal format, we observe that the specifications indicate a DS/SS-QPSK transmission with real spreading (Q-RS) which, according to Table 2-1, yields the best spectral efficiency. Two different options are indicated for the set of spreading signatures: *i*) a composite code set made of the orthogonal WH sequences overlaid by an extended PN for scrambling and cell/beam identification purposes, and *ii*) an extended version of the quasi-orthogonal Gold set, without overlaying. In addition the specifications envisage a variable spreading factor M in order to make the generator capable of supporting multi-rate transmissions. Since the modulation scheme is QPSK, the chip rate R_c of the useful signal is given by

$$ R_c = M\,R_s = n \cdot L\,R_s = n \cdot L\,\frac{R_b}{2}, \qquad (2.171) $$

where L is the signature code period, n represents the number of code periods within one symbol interval, R_s and R_b are the symbol rate and the bit rate, respectively. The permitted values of R_b, L, and R_c, evaluated according to (2.171), are shown in Table 2-3 (maximum chip rate $R_{c,max} = 2.048$ Mchip/s). Full programmability is supported as far as data rates, multiple access interference and additive noise are concerned. The aggregate CDMA signal, made of the useful signal and MAI, produced by the MUSIC generator is described by the model in (2.109). Eventually the MUSIC generator outputs an analog signal centered around the standard Intermediate Frequency (IF) $f_{IF} = 70$ MHz. The bandwidth occupancy is $B_{IF} = (1+\alpha)R_c$, where $\alpha = 0.22$ is the roll off factor of the SRRC chip pulse shaping filters and its maximum value turns out to be then $B_{IF} = (1+\alpha)R_{c,max} = 2.56$ MHz.

The receiver specifications are listed in Table 2-4. The CDMA signal format specifications are the very same as in the transmitter case. In addition, some specifications are reported for the step size of the adaptive detector and for the synchronization, namely the spreading code acquisition time and the *Mean Time to Lose Lock* (MTLL). The operating conditions, are expressed in terms of *Signal to Noise Ratio* (SNR) and/or *Bit Error Rate* (BER). The specifications also report the overall maximum SNR degradation allowed for the modem (implementation loss) owed to finite-precision arithmetic digital processing and imperfect receiver synchronization. Finally, the receiver specifications indicate two output formats: *i*) binary hard detected Non-Return to Zero (NRZ) data, and *ii*) soft output samples represented by 4 bits.

Table 2-2. Specifications of the MUSIC CDMA signal generator

Feature	Symbol	Specifications
Min. / max. channel data rate	R_b	4 / 128 kb/s
Chip rate	R_c	Programmable 0.128, 0.256, 0.512, 1.024, 2.048 Mchip/s accuracy: 1 Hz
Signature sequence type		Option I: WH (traffic code) + E-PN (overlay) Option II: E-Gold
Signature sequence period	L	32, 64, 128 chips
Modulation/spreading technique		QPSK / Balanced DS/SS with real spreading (i.e., single-code)
Spreading factor	M	$n \cdot L_c$, $n = 1, \ldots, 16$
Max. no. of interfering channels	N	L
Chip shaping	$g_T(t)$	Square root raised cosine with roll off 0.22
Random data generator stream		Programmable on each CDMA channel with disable capability
Interferers' delay with respect to the useful channel	τ_i	Programmable on each CDMA channel in the range $0 \div L$ chip intervals, resolution 0.1 chip interval
Interferers' phase offset with respect to the useful channel	ϑ_i	Programmable on each CDMA channel in the range $0 \div 360$ degrees, resolution 1 degree
Interferers' carrier frequency offset with respect to the useful channel	Δf_i	Programmable on each CDMA channel in the range ± 70 kHz, resolution 1 Hz
Useful channel to single interferer power ratio	C / I	Programmable on each CDMA channel in the range ± 10 dB , resolution 0.1 dB
IF carrier frequency	f_{IF}	70 MHz
Max. carrier frequency uncertainty on the useful channel		± 100 Hz
Output signal level		Programmable in the range $-30 \div -10$ dBm , step 5 dBm
SNR	E_b / N_0	Programmable in the range $-3 \div 30$ dB , step 1 dB

Table 2-3. Bit and chip rates of the MUSIC CDMA signal format

Rb [kb/s]	n	Rc [kchip/s] @ L = 32	Rc [kchip/s] @ L = 64	Rc [kchip/s] @ L = 128
4	1	–	128	256
	2	128	256	512
	4	256	512	1024
	8	512	1024	2048
	16	1024	2048	–
8	1	128s	256	512
	2	256	512	1024
	4	512	1024	2048
	8	1024	2048	–
	16	2048	–	–
16	1	256	512	1024
	2	512	1024	2048
	4	1024	2048	–
	8	2048	–	–
	16	–	–	–
32	1	512	1024	2048
	2	1024	2048	–
	4	2048	–	–
	8	–	–	–
	16	–	–	–
64	1	1024	2048	–
	2	2048	–	–
	4	–	–	–
	8	–	–	–
	16	–	–	–
128	1	2048	–	–
	2	–	–	–
	4	–	–	–
	8	–	–	–
	16	–	–	–

Table 2-4. Specifications of the MUSIC CDMA receiver

Feature	Symbol	Specifications
IF carrier frequency	f_{IF}	70 MHz
Max. carrier frequency uncertainty		± 100 Hz
Input signal dynamics		$-40 \div -10$ dBm
Max. power unbalance between traffic channels		± 6 dB
Min. input SNR	E_b / N_0	-1 dB
Min. / max. channel data rate	R_b	4 / 128 kb/s
Chip rate	R_c	Programmable 0.128, 0.256, 0.512, 1.024, 2.048 Mchip/s with accuracy: 1 Hz
Signature sequence type		Option I: WH (traffic code) + E-PN (overlay) – Option II: E-Gold
Signature sequence period	L	32, 64, 128 chips
Modulation/spreading technique		QPSK / Balanced DS/SS with real spreading (i.e., single-code)
Spreading factor	M	$n \cdot L, \ n = 1, \dots, 16$
Max no. of interfering channels	N	L
Chip shaping	$g_T(t)$	Square root raised cosine with roll off 0.22
Adaptive detector step size	γ_{BAID}	Programmable / adaptive with signal amplitude
Min. SNR for acquisition and tracking @ $E_s / N_0 = 0$ dB	$\left[\dfrac{E_c}{N_0}\right]_{min}$	-24 dB
Mean acquisition time @ $[E_c / N_0]_{min}$ and $E_s / N_0 = 0$ dB	$\overline{T}_{acq}$	< 4 sec
Acq. time with 99% @ $[E_c /(N_0 + I_0)]_{min}$ and $E_s / N_0 = 0$ dB	T_{acq}	< 8 sec
MTLL for code/phase tracking @ BER= $8 \cdot 10^{-2}$	$\overline{T}_{LL}$	$> 3 \cdot 10^4$ sec
Overall SNR degradation on AWGN w.r.t. theory @ $10^{-3} \le BER \le 8 \cdot 10^{-2}$		≤ 0.5 dB
Overall SNR degradation on AWGN w.r.t. floating point simulation, with ideal sync./EC-BAID configuration @ $10^{-3} \le BER \le 8 \cdot 10^{-2}$		≤ 1 dB
Baseband data output		Binary hard detected NRZ and 4 bit soft output, with NRZ clock signal

Chapter 3

DESIGN OF AN ALL DIGITAL CDMA RECEIVER

After introducing the fundamentals of spread spectrum signaling and CDMA we are now ready to delve deeply into the details of the system level design of a DSP-based CDMA receiver. We will follow a bottom up approach, starting from the multirate signal processing to be carried out in the front end section of the demodulator, down to the specific subtleties of the synchronization and signal detection algorithms. This will result in an overall receiver architecture whose description and simulated performance will be discussed in detail.

1. CDMA RECEIVER FRONT END

This Chapter contains a description of the main basic building blocks of a DSP-based, multirate digital CDMA receiver. Starting from the architecture of the digital downconversion stage in the MUSIC receiver's Front End (FE), it develops through a description of the signal interpolator to be used at the output of the Chip Matched Filter (CMF), and describes the detailed design of several sub-systems of an all digital CDMA receiver, with particular emphasis on the multi-rate CDMA demodulator and to the interference mitigation functionality.

1.1 Multi-Rate CDMA Signal

As already detailed in the previous Chapter, the signal at the output of the MUSIC signal generator is centered around Intermediate Frequency (IF) $f_{IF} = 70 \, \text{MHz}$ and has a bandwidth occupancy $B_{IF} = (1+\alpha)R_c$, where

$\alpha = 0.22$ is the roll off factor of the chip pulse shaping filter and R_c is the chip rate. The latter ranges from $R_{c,min} = 128$ to $R_{c,max} = 2048\,\text{kchip/s}$, so that he maximum signal bandwidth occupancy turns out to be $B_{IF,max} = (1 + \alpha)R_{c,max} = 2.56\,\text{MHz}$. A sketch of the bandwidth occupancy of the signal at the output of the signal generator is shown in Figure 3-1. Here, as well as in the subsequent drawings, the spectral components of the signal are represented as *asymmetric* with respect to f_{IF}. This is for illustrative purposes only, and is not intended to be a faithful illustration of the actual spectrum. The asymmetric spectral shape is helpful in identifying the positive and negative frequency components of the received IF signal spectrum during the baseband conversion process described in Section 1.3.

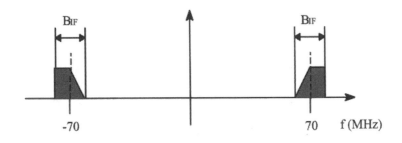

Figure 3-1. Spectrum of the IF signal.

1.2   Receiver Overall Architecture

The top level schematic of the CDMA receiver is shown in Figure 3-2. For ease of discussion, the different functions performed by the receiver can be partitioned as follows:

a)   Digital Downconversion Unit (DDU);
b)   Multirate Front End Unit (MRFEU);
c)   Linear Interpolation Unit (LIU);
d)   Code Timing Acquisition Unit (CTAU);
e)   Chip Clock Tracking Unit (CCTU);
f)   Automatic carrier Frequency Control Unit (AFCU);
g)   Signal Amplitude Control Unit (SACU);
h)   EC-BAID Unit, embedding Carrier Phase Recovery Unit (CPRU).

In addition the receiver also encompasses the Analog Signal Conditioning Unit (ASCU) which performs band pass limiting and amplitude control of the IF received signal prior to Analog to Digital Conversion (ADC), and a digital processing unit devoted to SNIR (Signal to

Noise plus Interference Ratio) estimation, denoted as SNIR Estimation Unit (SEU). However, we remark that the design issues related to the latter two additional units fall outside the scope of this book and therefore will not be addressed here. The interested reader can find more details about signal conditioning in reference [MUS01] and about SNIR estimation in references [Gilch], [Div98], [MUS01].

1.3　From Analog IF to Digital Baseband

Concerning ADC, two different approaches are possible: *asynchronous* and *synchronous signal sampling*.

In the first architecture the received signal is passed through an Anti-Alias Filter (AAF) and is then fed to the ADC. The latter is controlled by a *free running* oscillator, having no reference whatsoever with the clock of the data signal. The ADC sampling rate $1/T_s$ is usually higher than the symbol rate by an *over sampling factor* greater than 2. Timing correction is achieved by *interpolating* the samples available at the matched filter output according to the estimates of the timing error. Briefly speaking, the interpolator 're-synthesizes' those signal samples at the correct timing instants, that are in general not present in the digitized stream. More details on the interpolator sub-unit are reported in Section 2.2.

An alternative architecture involves *synchronous* signal sampling, and is particularly useful with high data rate modems where oversampling is too expensive or unfeasible altogether. Here, timing correction is accomplished through a feedback loop wherein a Timing Error Detector (TED) drives a Numerically Controlled Oscillator (NCO) that adaptively adjusts the clock of the ADC, so that the digitized samples are synchronous with the data clock.

In the MUSIC receiver, the ADC rate is not critical, and the decision was to resort to is asynchronous sampling. The received signal undergoes IF filtering and is passed to the ADC wherein it is sampled at rate f_s. The value of f_s is selected taking into account the following requirements:

i)　$f_s \geq 4\,R_{c,\max} = 4 \cdot 2.048\,\text{Mchip/s} = 8.192\,\text{MHz}$ (to yield at least four samples per chip);

ii)　$f_s = 2^n$ (to select from standard commercial quartz clocks);

iii)　$kf_s \geq 2f_{IF} + B_{IF}$, $(k-1)f_s < 2f_{IF} - B_{IF}$ (to ensure that the spectral replicas arising from ADC do not overlap).

According to *i*) and *ii*), we set $n = 14$ and $f_s = 16.384\,\text{MHz}$. Such value of f_s was also found to meet condition *iii*) for an IF bandwidth $B_{IF} = 2.56\,\text{MHz}$, with $k = 9$.

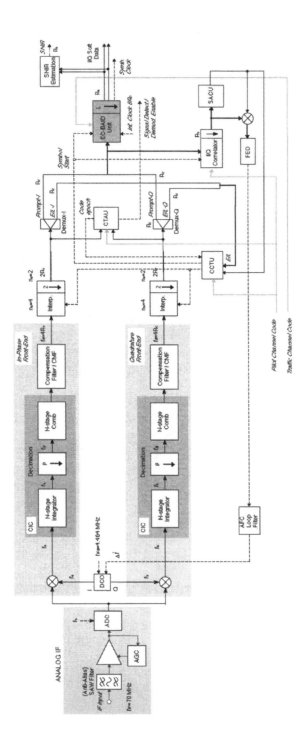

Figure 3-2. Top level schematic of the MUSIC receiver.

The spectrum of the resulting sampled signal is shown in Figure 3-3, where the replicas of the positive and negative frequency signal spectrum are centered around

$$f_k^{(+)} = 70 - k\, f_s, \tag{3.2.a}$$

$$f_k^{(-)} = -70 + k\, f_s, \tag{3.2.b}$$

respectively, with k an integer.

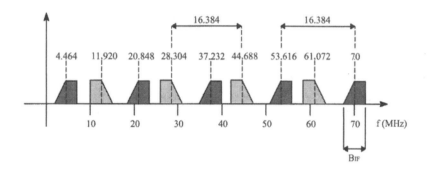

Figure 3-3. Spectrum of sampled IF signal after ADC.

Figure 3-4 zooms on the low frequency part of the spectrum, containing the following spectral replicas

$$f_4^{(+)} = 70 - 4\, f_s = 4.464 \text{ MHz}, \tag{3.3.a}$$

$$f_5^{(+)} = 70 - 5\, f_s = -11.920 \text{ MHz}, \tag{3.3.b}$$

$$f_4^{(-)} = -70 + 4\, f_s = -4.464 \text{ MHz}, \tag{3.3.c}$$

$$f_5^{(-)} = -70 + 5\, f_s = 11.920 \text{ MHz}. \tag{3.3.d}$$

The sampled signal is then digitally I/Q downconverted to baseband by a Digitally Controlled Oscillator (DCO) operating at the Digital IF (IFD) $f_{\text{IFD}} = 4.464$ MHz, as is sketched in Figure 3-5. The downconverted signal contains unwanted image spectra located at the frequencies $\pm f'$ and $\pm f''$ as follows

$$f_5^{(-)} - f_{\text{IFD}} = 7.456 \text{ MHz} = f', \tag{3.4.a}$$

$$f_4^{(+)} + f_{\text{IFD}} = 8.928 \text{ MHz} = f'', \tag{3.4.b}$$

$$f_5^{(+)} + f_{\text{IFD}} = -7.456 \text{ MHz} = -f', \tag{3.4.c}$$

$$f_4^{(-)} - f_{\text{IFD}} = -8.928 \text{ MHz} = -f''. \tag{3.4.d}$$

The unwanted image spectra located at $\pm f'$ and $\pm f''$ will be rejected by means of the (low pass) CMF. By normalizing with respect to the sampling frequency f_s, we also get the normalized locations of the spectral images to be rejected

$$\frac{f'}{f_s} = \frac{7.456}{16.384} \cong 0.455, \tag{3.5.a}$$

$$\frac{f''}{f_s} = \frac{8.928}{16.384} \cong 0.545. \tag{3.5.b}$$

The basic and conceptual schematic of the I/Q downconverter is shown in Figure 3-6, where the blocks labeled 'CIC' are in charge of decimation, as detailed in Section 1.4. Figure 3-7 shows the implementation of the scheme in Figure 3-6. (Digital) I/Q downconversion is implemented by resorting to a Direct Digital Synthesizer (DDS) that drives a couple of real multipliers. Particularly, the DDS is made up by a phase accumulator and a Look Up Table (LUT) storing the sine and cosine samples.

The phase accumulator performs numerical integration of the digital reference at frequency f_{IFD}, whose value is represented by the Frequency Control Word (FCW) on n_{FCW} bits, according to the following recursive equation computed at the clock rate f_s

$$\phi_k = \phi_{k-1} + 2\pi T_s f_{\text{IFD}}, \tag{3.6}$$

where $T_s = 1/f_s$ is the sampling interval. The phase accumulator, which is assumed to operate internally on n_{acc} bits, outputs a (sampled) ramp, represented by n_{pha} bits (with $n_{pha} < n_{acc}$), whose slope is determined by the digital FCW f_{IFD}. The n_{pha} most significant bits of the accumulator are then used to represent the phase ϕ_k that is passed to a LUT which performs sine/cosine generation and outputs the sequences $\sin(\phi_k)$ and $\cos(\phi_k)$, represented by n_{DCO} bits. To do this the word representing the phase ϕ_k on n_{pha} bits is used to address a ROM table containing N_{LUT} samples excerpted from a period of a cosinusoid, quantized by $2^{n_{\text{DCO}}}$ levels.

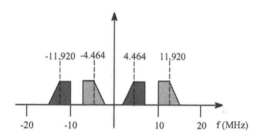

Figure 3-4. Particular of Figure 3-3.

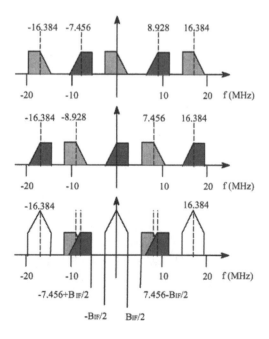

Figure 3-5. Digital downconversion to baseband.

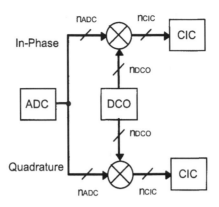

Figure 3-6. Basic block diagram of the I/Q digital downconversion to baseband.

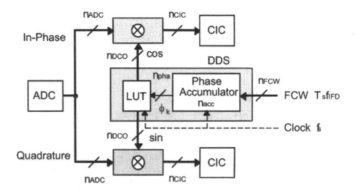

Figure 3-7. Implementation of the I/Q digital downconversion to baseband.

Let us focus on the issue of quantization of ϕ_k, $\sin(\phi_k)$ and $\cos(\phi_k)$, and let us determine the number of required bits. First of all, assume for the moment that the sine and cosine values have so many bits that the quantization of ϕ_k is significant only. If the phase is quantized by n_{pha} bits the unit circle is divided into $2^{n_{pha}}$ phased increments, and the quantization noise is uniformly distributed, so that the resulting RMS phase noise, measured in fractions of a cycle, is

$$\sigma_\phi = \frac{2^{-n_{pha}}}{\sqrt{12}}. \tag{3.7}$$

For instance, assuming $n_{pha} = 8$ bits, we have $\sigma_\phi = 1.1 \times 10^{-3}$ cycles, corresponding to an RMS value of 0.41 degrees, which seems adequately small for the detection of QPSK-modulated signals to bear no appreciable BER degradation. We remark however that accurate assessment of the actual BER penalty associated with n_{pha} can be carried out only by means of bit true computer simulations, as is done in Section 5.2.

Now, let us make the reverse assumption that ϕ_k is delivered with 'infinite' precision, and that significant quantization takes only place in the 'output' values of the sine and cosine functions. Denote also by q(ξ) the quantized versions of the infinite resolution value ξ. In this case, if the sine and cosine functions are quantized by n_{DCO} bits the received signal will be rotated by a phase amount given by

$$\phi_k' = \arctan\left\{\frac{q\left[\sin\left(\phi_k\right)\right]}{q\left[\cos\left(\phi_k\right)\right]}\right\}, \tag{3.8}$$

and, neglecting any amplitude fluctuation resulting from phase quantization, the quantization error is

$$\Delta\phi_k = \phi_k' - \phi_k. \tag{3.9}$$

Following the analysis in [Gar88], the RMS value of this error, measured in fractions of a cycle, is given by

$$\sigma_\phi = \frac{2^{-n_{DCO}}}{\pi\sqrt{12}}. \tag{3.10}$$

For $n_{DCO} = 8$ bits, we have $\sigma_\phi = 3.6 \times 10^{-4}$ cycles, corresponding to 0.13 degrees. From the two expressions above it turns out that the RMS quantization error on the sine/cosine function values is π times (i.e., about 4 times) larger than the RMS error caused by quantization of the phase (for the same word length). Therefore, if the sine/cosine LUT outputs are quantized by the same number of bits as the phase ϕ_k at the LUT input ($n_{DCO} = n_{pha}$), then the effect of quantization on the phase is dominant. Alternatively, if we let $n_{DCO} = n_{pha} - 2$ (i.e., the sine/cosine outputs are represented by a word shorter by two bits than the input phase word), then the LUT input and outputs provide nearly equal contributions to the net quantization noise.

In addition to the issue of quantization noise owed to I/Q and/or phase quantization, it is also necessary to determine the effect of finite length of a word representation in the DCO. The phase rotator described above acts also as a frequency translator according to the following recursive algorithm, which is inherently implemented by the DCO

$$\phi_k = \left(\phi_{k-1} + 2\pi T_s \, f_{\text{IFD}}\right) \bmod 2\pi, \tag{3.11}$$

where ϕ_k is the instantaneous phase used to correct the incoming signal. Scaling by 2π we obtain

$$\frac{\phi_k}{2\pi} = \left(\frac{\phi_{k-1}}{2\pi} + T_s \, f_{\text{IFD}}\right) \bmod 1. \tag{3.12}$$

Denoting now by ϑ_k the (normalized) phase expressed in cycles ($\vartheta_k = \phi_k / 2\pi, 0 \le \vartheta_k < 1$), we obtain

$$\vartheta_k = \left(\vartheta_{k-1} + T_s \, f_{\text{IFD}}\right) \bmod 1, \tag{3.13}$$

or

$$\vartheta_k = \left(k T_s \, f_{\text{IFD}}\right) \bmod 1, \tag{3.14}$$

which describes operation of the DCO. According to (3.13), at each sample time the FCW (i.e., the normalized frequency $T_s \, f_{\text{IFD}}$ represented by n_{FCW} bits) is added to the previous contents of the accumulator, which operates on n_{acc} bits. Overflow management of the accumulator is intrinsic to the modulo-1 operation, with the accumulator contents regarded as a binary fraction between 0 and 1 cycle. The exact value of the (normalized) frequency $f_{\text{IFD}} T_s$ is provided (at symbol time) by the Frequency Error Detector (FED). Phase resolution can be made as fine as desired by lengthening the accumulator word length, i.e., by increasing n_{acc}. The Least Significant Bit (LSB) of the FCW ($T_s \, f_{\text{IFD}}$) must be consistent with the LSB of the accumulator. If there are n_{acc} bits in the accumulator, the phase is quantized into increments of $2^{-n_{acc}}$ cycle. Usually only the n_{pha} Most Significant Bits (MSBs) of the accumulator (with $n_{pha} \le n_{acc}$) are read out to the subsequent sine/cosine LUT. The smallest frequency increment is

$$\Delta f = \frac{2^{-n_{acc}}}{T_s} = \frac{f_s}{2^{n_{acc}}} \text{ Hz}, \tag{3.15}$$

so improved frequency resolution can be obtained by lengthening the accumulator (and the relevant FCW).

In many practical applications a typical choice for the accumulator word length is $n_{acc} = 32$, so as to provide a phase resolution

$$\Delta\vartheta = 2^{-n_{acc}} = 2^{-32} \cong 2.33 \times 10^{-10} \text{ cycles,} \tag{3.16}$$

or, equivalently,

$$\Delta\phi = 2^{-n_{acc}} \cdot 2\pi = 2^{-32} \cdot 2\pi \cong 1.46 \times 10^{-9} \text{ rad,} \tag{3.17.a}$$

$$\Delta\phi = 2^{-n_{acc}} \times 360 = 2^{-32} \times 360 \cong 8.38 \times 10^{-8} \text{ deg.} \tag{3.17.b}$$

Concerning the frequency resolution we have

$$\Delta f = \frac{f_s}{2^{n_{acc}}} = \frac{16.384 \text{ MHz}}{2^{32}} = 3.815 \times 10^{-3} \text{ Hz} \tag{3.18}$$

and

$$\frac{\Delta f}{f_s} = \frac{1}{2^{n_{acc}}} = 2.33 \times 10^{-10}, \tag{3.19}$$

which is unnecessarily accurate for our purpose. By relaxing the requirements about the resolution, we may consider a smaller accumulator based on $n_{acc} = 8$ bits, thus obtaining

$$\Delta\vartheta = 2^{-n_{acc}} = 2^{-8} \cong 3.91 \times 10^{-3} \text{ cycles} \tag{3.20}$$

$$\Delta\phi = 2^{-n_{acc}} \times 2\pi = 2^{-8} \times 2\pi \cong 3.91 \times 10^{-3} \times 2\pi \cong 2.45 \times 10^{-2} \text{ rad,} \tag{3.21.a}$$

$$\Delta\phi = 2^{-n_{acc}} \times 360 = 2^{-8} \times 360 \cong 3.91 \times 10^{-3} \times 360 \cong 1.41 \text{ deg,} \tag{3.21.b}$$

$$\Delta f = \frac{f_s}{2^{n_{acc}}} = \frac{16.384 \text{ MHz}}{2^{8}} = 64 \text{ kHz,} \tag{3.22}$$

$$\frac{\Delta f}{f_s} = \frac{1}{2^{n_{acc}}} = 3.91 \times 10^{-3}. \tag{3.23}$$

The feasibility of an 8 bit accumulator will be verified later by means of extensive bit true computer simulations. The phase at the accumulator output can not have $n_{pha} = n_{acc}$ because of speed limitations in the access to the LUT, therefore only the n_{pha} MSBs of the accumulator (with $n_{pha} \leq n_{acc}$) are read out to the following sine/cosine LUT. According to the results highlighted above, it is common practice to choose $n_{pha} = n_{DCO} + 2$ so as to limit the truncation error. We assume that signal degradation is negligible if the digital phase rotation of the samples is carried out by using 8 bit sine/cosine coefficients ($n_{DCO} = 8$ bits). This assumption seems rather conservative and shall be verified, and possibly relaxed, by means of bit true simulations. Under this hypothesis we have $n_{pha} = 10$ bits, and we can address a number of $2^{n_{pha}} = 2^{10} = 1024$ different phases. Considering the symmetries of the sine and cosine functions, we can reduce the number of elements stored in the LUT. Actually, it is sufficient to store in the LUT only the samples (represented by 8 bits) of the sine in the interval $0-\pi/2$ (corresponding to 1/4 of a cycle). Recalling that the phase resolution is 1024 samples per cycle, the LUT must contain only $N_{LUT} = 1024/4 = 256$ values. A pictorial diagram of the LUT architecture is shown in Figure 3-8.

1.4   Decimation and Chip Matched Filtering

The received signal is sampled at the rate $f_s = 1/T_s = 16.384$ Msample/s, and the number of *samples per chip* is therefore given by

$$\nu_s = \frac{f_s}{R_c} = \frac{16.384}{R_c}, \tag{3.24}$$

whose possible values for the various chip rates are reported in Table 3-2.

Notice that chip matched filtering and the subsequent digital processing operations (noticeably chip timing synchronization) require no more than 2 to 4 samples per chip interval. According to (3.24) the signal turns out to be significantly oversampled for the lowest bit rates, and this would cause the CMF to bear excessive complexity owing to the huge number of taps required. Therefore *decimation* is in order so as to achieve the target sampling rate of $n_s = 4$ sample/chip. The decimation factor to be applied is

$$\rho = \frac{\nu_s}{n_s} = \frac{f_s}{n_s R_c} = \frac{16.384}{4 R_c} = \frac{4.096}{R_c} \tag{3.25}$$

and its numerical values are reported in Table 3-1.

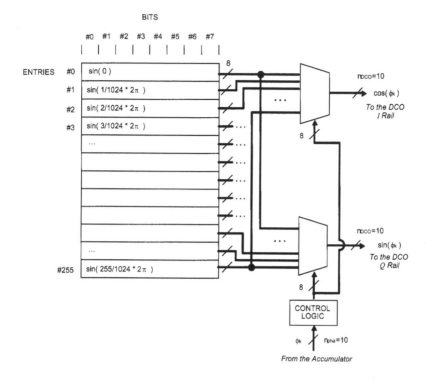

Figure 3-8. LUT architecture.

A signal sampled at rate f_s is decimated by a factor ρ by taking the first sample out of a group of ρ samples. The decimated signal has frequency $f_d = f_s/\rho$. From a spectral point of view this implies that the spectral replicas, which in the original signal were located on frequencies $m\ f_s$ (with m an integer), are moved to $mf_d = m\ f_s/\rho$. Unfortunately this operation causes aliasing of the wideband noise spectrum, thus reducing the Signal to Noise Ratio (SNR). Figure 3-9 shows this phenomenon for a decimation factor $\rho = 2$.

Table 3-1. Chip Rates (R_c), Samples per Chip (v_s), Decimation Factors (ρ), and Decimated Frequencies (f_d).

R_c	v_s	ρ	f_d
0.128 Mchip/s	128 sample/chip	32	0.512 Msample/s
0.256 Mchip/s	64 sample/chip	16	1.024 Msample/s
0.512 Mchip/s	32 sample/chip	8	2.048 Msample/s
1.024 Mchip/s	16 sample/chip	4	4.096 Msample/s
2.048 Mchip/s	8 sample/chip	2	8.192 Msample/s

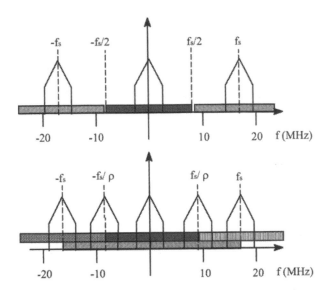

Figure 3-9. Aliasing of white noise spectrum.

When decimating a signal we have to ensure by appropriate pre-filtering that no aliasing will take place. A low complexity solution for jointly performing pre-filtering and decimation is the so called Cascaded Integrator and Comb (CIC) architecture [Hog81], depicted in Figure 3-10 and described hereafter. The integer N represents the order of the CIC filter, while the integer M is termed differential delay.

From Figure 3-10 the transfer function of the N stage cascaded integrator section, operating at sample rate f_s, is

$$H_I(z) = \left(\frac{1}{1-z^{-1}}\right)^N ,$$ (3.26)

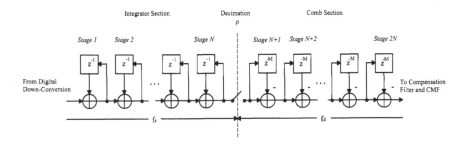

Figure 3-10. CIC decimation filter.

After the integrator section the sample stream is decimated by a factor ρ, down to rate $f_d = 1/T_d$ given by

$$f_d = \frac{f_s}{\rho} = \frac{16.384}{\rho} = \frac{16.384}{4.096} R_c = 4 R_c. \qquad (3.27)$$

The numerical values of the decimated rate f_d envisaged in the framework of the MUSIC project are reported in the last column of Table 3-1. Finally, the decimated samples are filtered by the N stage cascaded comb section, operating at rate f_d, and whose transfer function is

$$H_C'(z) = \left(1 - z^{-M}\right)^N. \qquad (3.28)$$

The transfer function of the comb section can be re-written with respect to the input sample rate f_s as

$$H_C(z) = \left(1 - z^{-\rho M}\right)^N. \qquad (3.29)$$

The resulting overall transfer function, expressed with respect to the input sample rate f_s, is then

$$H(z) = \left(\frac{1 - z^{-\rho M}}{1 - z^{-1}}\right)^N = \left(\sum_{k=0}^{\rho M - 1} z^{-k}\right)^N. \qquad (3.30)$$

The frequency response of the CIC filter is therefore

$$H(f) = e^{-j\pi f T_s (\rho M - 1) N} \left[\frac{\sin(\pi f T_s \rho M)}{\sin(\pi f T_s)}\right]^N, \qquad (3.31)$$

where $f_s = 1/T_s$. The amplitude response is

$$|H(f)| = \frac{\left|\sin\left(\pi \rho M \dfrac{f}{f_s}\right)\right|^N}{\left|\sin\left(\pi \dfrac{f}{f_s}\right)\right|^N} = \left|M\rho \frac{\text{sinc}\left(\rho M \dfrac{f}{f_s}\right)}{\text{sinc}\left(\dfrac{f}{f_s}\right)}\right|^N, \qquad (3.32)$$

or equivalently, recalling that $f_d = f_s / \rho$,

$$
\left| H(f) \right| = \frac{\left| \sin\left(\pi M \dfrac{f}{f_d} \right) \right|^{N}}{\left| \sin\left(\pi \dfrac{f}{\rho f_d} \right) \right|^{N}} = \left| M\rho \dfrac{\mathrm{sinc}\left(M \dfrac{f}{f_d} \right)}{\mathrm{sinc}\left(\dfrac{f}{\rho f_d} \right)} \right|^{N} . \tag{3.33}
$$

We observe now that

$$
H(0) = \left(M\rho \right)^{N}, \tag{3.34}
$$

and therefore the normalized frequency response is

$$
G(f) = \frac{H(f)}{H(0)} = e^{-j\pi f T_s (\rho M - 1)N} \left[\frac{1}{M\rho} \frac{\sin\left(\pi f T_s \rho M \right)}{\sin\left(\pi f T_s \right)} \right]^{N}, \tag{3.35}
$$

so $G(0) = 1$. From the equations above it is apparent that the CIC filter is equivalent to a cascade of N 'moving average' Finite Impulse Response (FIR) stages, each operating at the input sample rate f_s and having transfer function

$$
S(z) = \sum_{k=0}^{\rho M - 1} z^{-k} \tag{3.36}
$$

followed by a decimator which reduces the sample rate by a factor ρ, as shown in Figure 3-11, where the transfer functions $H(z)$ and $S(z)$ are also indicated.

Implementation of the CIC in this second form would be much more complex than the original one, and this motivates the introduction of such a circuit. In analytical computations, it is sometimes more convenient to refer to the latter equivalent moving average model.

The baseband bandwidth occupancy of the useful signal is now

$$
B_{\mathrm{BB}} = \frac{B_{\mathrm{IF}}}{2} = \frac{(1 + \alpha) R_c}{2} = 0.61\, R_c , \tag{3.37}
$$

and the baseband occupancy normalized with respect to the decimated frequency f_d is given by

$$\beta_{BB} = \frac{B_{BB}}{f_d} = \frac{0.61\,R_c}{4\,R_c} = 0.1525\,, \tag{3.38}$$

which does not depend on the chip rate. Figure 3-12 shows the generic frequency response $G(f)$, as compared with the various (wanted and unwanted) spectral components of the received signal.

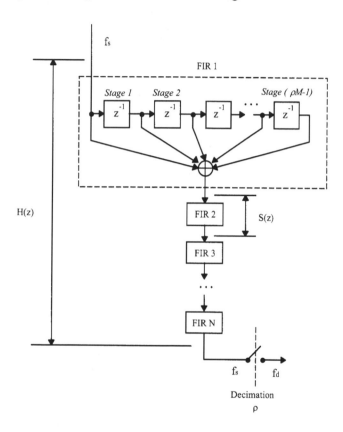

Figure 3-11. Equivalent model for the CIC decimation filter.

It is apparent that the amplitude response $G(f)$ of the CIC filter is not flat within the useful signal bandwidth, and therefore some compensation, by means of a subsequent equalizer, is required in order to minimize signal distortion. We also see that the particular value of the decimation factor ρ determines the location of the frequency response's nulls at the frequencies $mf_d = m\,f_s/\rho$. Such nulls reveal crucial for rejecting those spectral components that, owing to the decimation, are moved into the useful signal baseband. The differential delay M causes the appearance of intermediate nulls in between two adjacent nulls at mf_d. These additional nulls are of little utility and do not significantly increase the alias rejection capability of the CIC filter. This feature is highlighted in Figure 3-12, where the case $M=1$ (dashed line) is compared with the case $M=2$ (solid thick line). Actually, an increase of M does not yield any improvement in the rejection of the unwanted spectral components, while it requires an increase in the storage capability of the CIC filter. Therefore according to [Hog81] and [Har97] we will restrict our attention in the sequel to the case $M=1$.

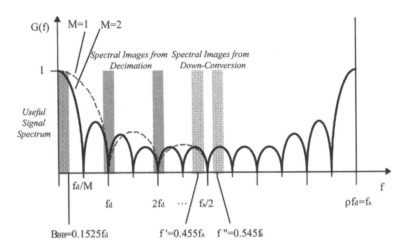

Figure 3-12. Generic normalized frequency response of the CIC decimation filter.

The order N of the CIC filter determines the sharpness of the notches at mf_d and the amplitude of the relevant sidelobes, therefore it must be carefully selected, taking into account the required attenuation of the unwanted spectral components. Assuming that a white noise process is superimposed on the signal at the CIC filter input, the shape of the frequency response $G(f)$ is proportional to the amplitude spectral density (i.e., the square root of the power spectral density) of the noise process at the output of the CIC, prior to decimation. Decimation causes the (normalized) amplitude spectral density $G(f)$ to be translated onto $mf_d = m\,f_s\,/\rho$. As a consequence the useful signal spectrum will suffer from aliasing caused by the lobes of the spectral replicas, as clarified in Figure 3-13.

The total contribution of the aliasing spectral replicas, that we call *alias profile* [Har97], is made of the contribution of ρ terms, and is bounded from above by the function

$$A(f) = \sum_{\substack{k=-\rho/2 \\ k\neq 0}}^{\rho/2} \left| G\left(f - kf_d\right) \right|.$$

(3.39)

The parameter N therefore keeps the alias profile $A(f)$ as low as possible within the useful signal's bandwidth B_{BB}.

Figure 3-14 shows the frequency response $G(f)$ for the different decimation ratios ρ in Table 3.2, for $M = 1$, and $N = 4$, while Figure 3-15 reports $G(f)$ for different orders of the filter N, for $M = 1$, and $\rho = 8$. In both the figures $G(f)$ is plotted versus the normalized frequency f / f_s.

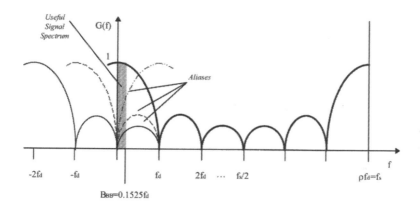

Figure 3-13. Aliasing effect of the CIC filter caused by decimation.

As already mentioned, the spectrum of the signal at the output of the CIC filter, at the decimated rate f_d, suffers from amplitude distortion, owing to the non-constant frequency response $H(f)$ (or, equivalently, $G(f)$). This calls for the use of a *compensation filter* (also termed *equalizer*) having a frequency response $H_{eq}(f)$ given by

$$H_{eq}(f) = \left[\frac{\sin(\pi f / \rho f_d)}{\sin(\pi M f / f_d)} \right]^N \qquad (3.40)$$

such that

$$\left| H_{eq}(f) H(f) \right| = 1. \qquad (3.41)$$

We will consider the compensation filter $G_{eq}(f)$ for the normalized frequency response $G(f)$, that is

$$G_{eq}(f) = \left[M\rho \frac{\sin\left(\pi \frac{f}{\rho f_d}\right)}{\sin\left(\pi M \frac{f}{f_d}\right)} \right]^N \tag{3.42}$$

such that

$$\left| G_{eq}(f) G(f) \right| = 1 , \tag{3.43}$$

with $G_{eq}(0) = 1$. The ideal (3.42) has Infinite Inpulse Response (IIR), and is approximated in our implementation as an FIR filter with N_{eq} taps and coefficients $g_k^{(eq)}$, where $k = 0,\dots,(N_{eq}-1)$. The actual frequency response of the equalizer $\hat{G}_{eq}(f)$ is

$$\hat{G}_{eq}(f) = \sum_{k=0}^{N_{eq}-1} g_k^{(eq)} \, e^{-j2\pi f k T_d} , \tag{3.44}$$

where T_d represents the sampling interval after decimation. Owing to the truncation of the impulse response we only have $\hat{G}_{eq}(f) \cong G_{eq}(f)$, and

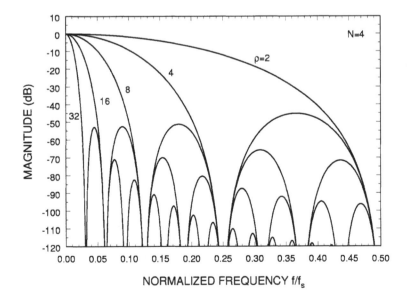

Figure 3-14. Frequency response of the CIC filter, $M = 1$, $N = 4$.

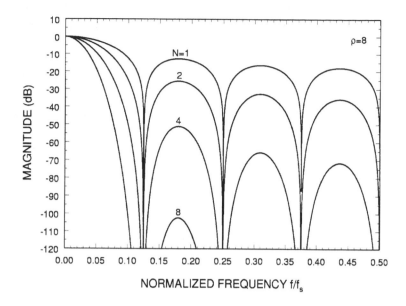

Figure 3-15. Frequency response of the CIC filter, $M = 1$, $\rho = 8$.

$$\widehat{G}_{eq}(0) = \sum_{k=0}^{N_{eq}-1} g_k^{(eq)} \neq 1. \tag{3.45}$$

Therefore we consider a re-normalized compensation filter $\widehat{G}'_{eq}(f)$, defined as

$$\widehat{G}'_{eq}(f) = \frac{\widehat{G}_{eq}(f)}{\widehat{G}_{eq}(0)} \tag{3.46}$$

such that $\widehat{G}'_{eq}(0) = 1$. The compensation filter $\widehat{G}_{eq}(f)$ can be synthesized according to the technique described in [Sam88], where a suitably modified version of the Parks–McClellan algorithm [McC73] for the design of equi-ripple FIR filters is used. The algorithm inputs are the length N_{eq} of the FIR impulse response and the bandwidth $\sigma_{n_h}^2 = E\{n_h^2(m)\} = N_0 / T_c$ with maximum flatness, after equalization. After some preliminary tries we set $N_{eq} = 17$, in order to reduce the complexity of implementation, and $B_F = 0.35 f_d$, so as to minimize amplitude distortion (in band ripple) on the signal bandwidth $B_{BB} = 0.1525 f_d$. As is apparent from the definition of $H_{eq}(f)$ and its related expressions, the frequency response of the equalizer depends on the decimation factor ρ. As a consequence the set of the coefficients of the compensation FIR filter must be computed and stored for every value of ρ, and the filter must be initialized by loading the coefficients $g_k^{(eq)}$ every time ρ is changed.

Figures 3-16 and 3-17 show the frequency response of the CIC compared with the alias profile, either uncompensated (dashed curves) or with compensation (solid curves), obtained for $\rho = 32$, with $N = 4$, $M = 1$. The effectiveness of the equalizer in flattening the frequency response up to $0.35 \, f_d$ is apparent. Also, with the parameters specified above, alias suppression within the useful bandwidth ($B_{BB} = 0.1525 \, f_d$) turns out to be higher than 45 dB.

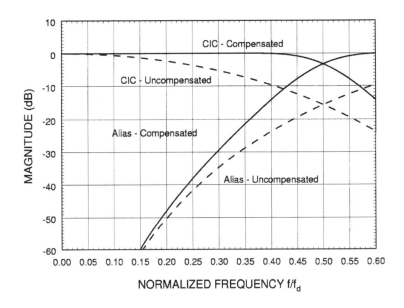

Figure 3-16. Frequency response of the CIC filter and alias profile, with (solid) and without (dashed) compensation, $M = 1$, $N = 4$, $\rho = 32$.

After equalization, filtering matched to the chip pulse takes place. The CMF is implemented with an FIR filter with N_{CMF} taps, approximating the ideal *Nyquist's Square Root Raised Cosine* (SRRC) frequency response

$$G_R(f) = \sqrt{\frac{G_N(f)}{T_c}}, \qquad (3.47)$$

where $G_N(f)$ is the *Nyquist's Raised Cosine* (RC) pulse spectrum with $G_N(0) = T_c$, and roll off factor $\alpha = 0.22$. Preliminary investigation about truncation effects in the CMF, carried out by computer simulation, demonstrated that the performance degradation is negligible if the SRRC impulse response is truncated (rectangular window) to $L = 8$ chip intervals. The overall length of the CMF impulse response must be at least

$$N_{CMF} = L \cdot n_s + 1 = 8 \cdot 4 + 1 = 33 \text{ samples}. \tag{3.48}$$

Considering the symmetry of FIR impulse response, the number of filter coefficients to be stored is

$$N'_{CMF} = \frac{(N_{CMF} - 1)}{2} + 1 = 17 \text{ coefficients}. \tag{3.49}$$

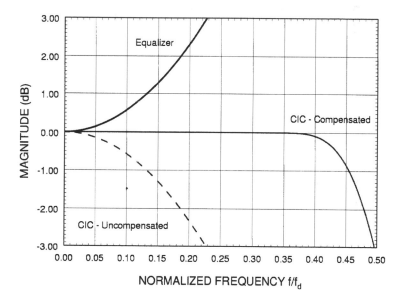

Figure 3-17. Frequency response of the CIC filter (dashed line), compensation filter (solid thick), and overall compensated response (solid thin), M = 1, N = 4, ρ = 32.

Integration of the compensation filter and the CMF into a single FIR filter was also considered.

However, the design of a single equivalent filter revealed quite a critical task. In particular, the resulting filter exhibited intolerable distortion on the slope of the frequency response. The consequence was that the two filters were implemented separately.

The resulting architecture of the front end of the MUSIC receiver is shown in Figure 3-18.

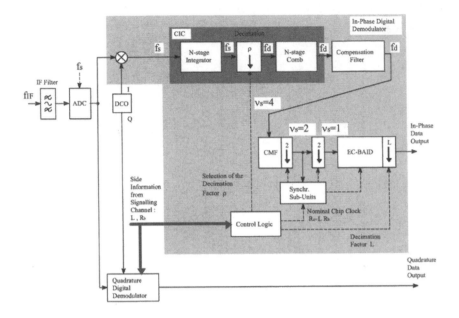

Figure 3-18. Architecture of the MUSIC receiver with the Multi-Rate Font-End.

2.   CDMA RECEIVER SYNCHRONIZATION

This Section tackles the issue of synchronization in a CDMA receiver, starting from a few general concepts, down to the particular design solutions adopted and implemented in the MUSIC receiver.

2.1   Timing Synchronization

During start up, and before chip timing tracking is started, the receiver has to decide whether the intended user m is transmitting, and, in the case he/she actually is, coarsely estimate the signal delay τ_m to initiate fine chip time tracking and data detection.

2.1.1   Code Timing Acquisition

Consider now the issue of code timing acquisition. In most cases this task is carried out by processing the so called *pilot signal*. This is a common CDMA channel in the forward link or a dedicated CDMA channel in the uplink, that is transmitted time and phase synchronous with the useful traffic signal(s), and whose data modulation is either absent or known *a priori*.

The pilot signature code sometimes belongs to the same orthogonal set (i.e., the Walsh–Hadamard set) as those used for the traffic channels. In this case, it is common practice to select as the signature of the pilot signal the 'all 1' sequence, i.e., the first row of the Walsh–Hadamard matrix.

However, in some cases it may be expedient to use a signature belonging to a *different* set (hence non-orthogonal) in order to avoid false locks owed to high off sync cross-correlation values of the WH sequences.

This issue will be addressed later when dealing with numerical results. Also the pilot signal is usually transmitted with a power level significantly higher than the traffic channel(s) (the so called *pilot power margin* or *P/C* ratio) to further ease acquisition and tracking.

As is discussed in [Syn98], conventional *serial acquisition* circuits are remarkably simple, but entail a time consuming process, leading to an *a priori* unknown acquisition time.

Therefore we have stuck to the *parallel acquisition* circuit for QPSK whose scheme is depicted in Figure 3-19. The design parameters of such a circuit are the value of the normalized threshold λ, and the length W of the *post-correlation smoothing window*. We shall not discuss here the impact of such parameters on acquisition performance, since this issue is well known from ordinary detection theory.

Implementation of the CTAU directly follows the general scheme in Figure 3-19, and is summarized in Figure 3-20 [De98d], [De98e]. The CTAU receives the stream of complex-valued samples at rate $2R_c$ (two samples per chip) at the output of the LIU.

Such an I/Q signal is processed by a couple of filters matched to the spreading code (this operation is also addressed to as the *sliding correlation* of the received signal with the local replica code). Notice that in Figure 3-20 the front end features *two* correlators because modulation is QPSK with *real* spreading (i.e., it uses a single code). Also the circuit in Figure 3-19 assumes a correlation length (the impulse response length of the front end FIR filters) equal to *one symbol span*, just as in the conventional despreader for data detection.

On the other hand, if we assume an *unmodulated* pilot there is no need in principle to limit the correlation length to one symbol (as, in contrast, is needed when data modulation is present). We have thus a further design parameter represented by the length of the correlation window.

For convenience we will investigate configurations encompassing a correlation time equal to an integer number, say M, of symbol periods (coherent correlation length).

The correlator outputs, again at the rate $2R_c$, are subsequently squared and combined so as to remove carrier phase errors. Parallelization takes place on the signal at the output of the combiner, still running at twice the chip rate. By parallelizing we obtain a $2L$-dimensional vector signal running at symbol time, whose components thus represent the (squared) correlations of the received signal with the locally generated sync reference signature code, for all of the possible 1/2-chip relative shifts of the start epoch of the latter.

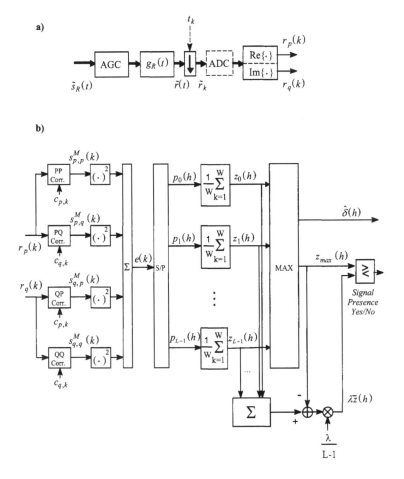

Figure 3-19. Parallel Code Acquisition Circuit.

After (parallel) smoothing on the observation window of length W symbols we obtain the sufficient statistics to perform signal recognition and ML estimation of the signature code initial phase. In particular, the maximum among all of the components is assumed to be the one bearing the 'correct' code phase. The CTAU broadcasts such information (denoted to as *code phase*) to all of the signature code generators that are implemented in the receiver (EC-BAID, CCTU, SACU etc.) either for traffic or for sync reference code generation. As is seen, this acquisition device also features an adaptive estimator of the noise plus interference level that is used to detect presence of the intended sync signal. The circuit also provides an information bit which indicates the presence, or the absence, of the pilot signal.

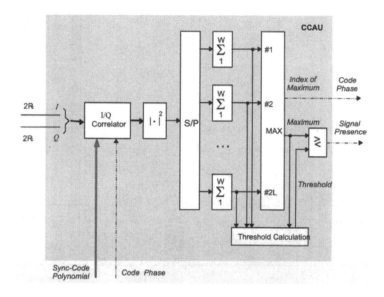

Figure 3-20. Block diagram of the CTAU.

In our design we set the CTAU parameters so as to obtain:

i)   probability of False (signal) Detection (P_{FD}) lower than 0.001;
ii)   probability of Missed (signal) Detection (P_{MD}) lower than 0.001;
iii)   probability of Wrong (code phase) Acquisition (P_{WA}) lower than 0.001.

Such probabilities are sufficiently low so as to enable one to disregard the influence on system performance of 'bad' events (i.e., acquisition takes place with approximately unit probability, and always takes place on the correct code phase). Considering the post-correlation smoothing period and the coherent correlation time, the total acquisition time is

$$T_{acq} = W \cdot M \cdot T_s = W \cdot M \cdot L \cdot T_c = \frac{W \cdot M \cdot L}{R_c} . \qquad (3.50)$$

The worst case corresponds to the lowest chip rate R_c = 0.128 Mchip/s, so that the acquisition time is bounded from above by

$$T_{acq} \leq \frac{W \cdot M \cdot L}{0.128} \, \mu s \qquad (3.51)$$

Recalling the requirement of the average acquisition time $\bar{T}_{acq} = 4 \, s$ in the project specifications, we have

$$\frac{W \cdot M \cdot L}{0.128} \, \mu s \leq 4 \times 10^6 \, \mu s, \qquad (3.52)$$

which gives

$$W \cdot M \cdot L \leq 512000. \qquad (3.53)$$

Table 3-2 reports the upper bounds of the product $W \cdot M$, referred to as *latency*, for the different code lengths.

Table 3-2. Upper bounds of the product *WM* (latency) for the CTAU.

L	$(W \cdot M)_{max}$
32	16000
64	8000
128	4000

2.1.2   Chip Timing Tracking

Once signal detection and coarse code timing acquisition have been successfully completed, chip timing tracking is started.

The unit in charge of fine chip time recovery is the CCTU, and is based on a non-coherent non-data aided closed loop tracker that closely follows the architecture outlined in [DeG93]. In this respect Figure 3-21 shows the integrated CCTU/LIU.

As apparent from the figure the outputs of both I and Q interpolators, running at the rate $2R_c$, are demultiplexed in two low rate (R_c) signals by two demultiplexers. The first signal is obtained collecting those samples taken (interpolated) at the optimum sampling instants, and are therefore referred to as *prompt* (or *on time*) samples. The other stream is made of the samples in between two consecutive prompt samples, and are therefore addressed to as *Early/Late* (E/L) samples. The prompt samples are used by the EC-BAID for data detection and by the Frequency Error Detector (FED) for fine carrier tuning, while the E/L samples are used by the CCTU for fine chip clock recovery.

More in detail the CCTU is made of a Chip timing Error Detector (CED) that operates on the E/L samples and an update unit which recursively updates the integer delay and the fractional epoch input to the LIU. The CED (shown in Figure 3-22) is the traditional non-coherent E/L correlator with time offset equal to one chip and full symbol correlation. The update rate of the CCTU output parameters is thus equal to the symbol rate (one CED output per symbol time). In order to ease clock tracking the CCTU performs correlation of the received samples with a local replica of the pilot signature code. Just to reduce implementation complexity, the squared amplitude nonlinearity of traditional E/L CEDs is replaced by a simpler amplitude nonlinearity. The relevant performance difference was shown to be negligible by simulation. The CED output signal is finally scaled by an amplitude control signal provided by the SACU, resulting in the arrangement sketched in Figure 3-22.

The CCTU is also equipped with the Lock Indicator shown in Figure 3-23 which signals completion of the timing lock procedure. The lock signal is obtained through a number of steps: first, the CED output ε_k is low pass filtered with the same bandwidth as the CCTU loop bandwidth; then the filter output is rectified; and finally the lock condition is tested through a comparator with hysteresis. The latter feature prevents possible sequences of repeated lock/unlock indication in a noisy environment.

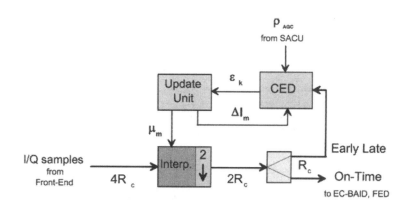

Figure 3-21. CCTU/LIU Architecture.

The initial state of the smoothing filter, as well as the comparator thresholds, are set according to the average value of the decision variable $|E_k|$, the so called *M curve*, that is shown in Figure 3-24.

The initial value of the detector status (i.e., of the smoothing filter output) E_0 has to be set taking into account the diverse initial sampling errors that may occur.

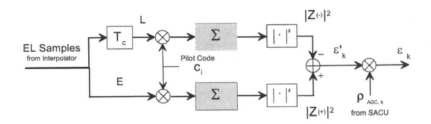

Figure 3-22. CED outline.

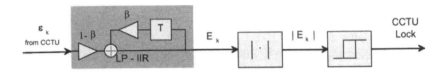

Figure 3-23. Lock Indicator.

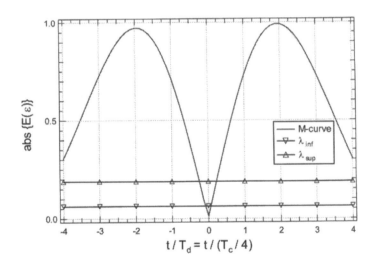

Figure 3-24. Lock Detector Characteristics (M curve).

The worst case is $\tau_0 = T_c / 4$ which corresponds to an average CED output equal to 0.5. If we want to signal loop lock when the timing error is smaller than or equal to 5% of a chip interval, the 'low' (or *inferior*) threshold must be roughly $\lambda_{inf} = 0.0625$ as shown in Figure 3-24 (dash–dot line). Also, if we want to signal loss of lock when the error is greater than

12.5% of a chip interval we have to set the 'high' (or *superior*) threshold to $\lambda_{\text{sup}} = 0.1875$.

Unfortunately, setting the smoothing filter onto a *positive* value fails when the initial timing error is *negative*. To attain symmetry in this respect, it is expedient to resort to the modified lock detection structure shown in Figure 3-25, where the two smoothing filters are initialized at the two symmetric values $-E_0$ and E_0, respectively. In so doing, the behavior of the detector will be always symmetric.

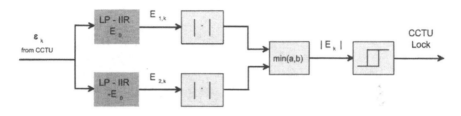

Figure 3-25. Modified Lock Detector.

2.2   Interpolation

The output $x(kT_d)$ of the I (or Q) CMF (with k running at the decimated sample rate $f_d = f_s/\rho = 4R_c$) are input to an interpolator (LIU) which provides the strobes for signal detection and synchronization (addressed to as prompt and E/L samples, see Section 2.1.2). Very accurate interpolation for band limited signals is in general provided by a third-order polynomial interpolator. In our case the digital signal bears a relatively high oversampling ratio (i.e., $f_d/R_c = 4$ samples per chip interval), so that a simpler linear (first-order) interpolator ensures sufficient accuracy.

In order to compensate for the drift between the free running clock of the receiver ADC and the actual chip clock of the received signal, each interpolator is controlled by an estimate of the (time varying) code timing delay provided by the CCTU. The signal $x(kT_d)$ running at $f_d = 1/T_d$ is then interpolated so as to provide a decimated signal at twice the chip rate $2R_c = 2/T_c$. During the generic m th symbol interval, we will have therefore $2L$ sampling epochs $t_{m,n}$ for any interpolator such that

$$t_{m,n+1} = t_{m,n} + \frac{T_c}{2}, \tag{3.54}$$

where m runs at the symbol rate $1/T_s$, n runs at twice the chip rate $2R_c = 2/T_c$, with $0 \le n \le 2L - 2$, and where $t_{m,0}$ is provided and renewed at each symbol interval by the CCTU. The initial sampling epoch $t_{m,0}$ in each symbol interval is in fact updated once per symbol interval as follows

$$t_{m,0} = mT_s + \tau_m ,$$ (3.55)

where τ_m is the control signal provided by the CCTU according to the following recursive equation

$$\tau_{m+1} = \tau_m - \gamma e_m .$$ (3.56)

In (3.56) e_m is the error signal provided by the CED of the CCTU (see Figure 3-21), and γ is the update step of the algorithm, the so called step size which in the following will be also referred as γ_{CCTU}. The stepsize γ must be set as a reasonable trade off between acquisition time and steady state jitter performance. From (3.55) we obtain

$$\tau_m = t_{m,0} - mT_s ,$$ (3.57)

$$\tau_{m+1} = t_{m+1,0} - (m+1)T_s ,$$ (3.58)

and substituting (3.57)–(3.58) into (3.56) we obtain

$$t_{m+1,0} - (m+1)T_s = t_{m,0} - mT_s - \gamma e_m$$ (3.59)

and

$$t_{m+1,0} = t_{m,0} + T_s - \gamma e_m .$$ (3.60)

The sequence $t_{m,0}$ needs, however, further processing in order to produce a control signal for the interpolator. From (3.69) it is seen that even if the update term γe_m takes on small values (say, fractions of a symbol interval T_s), the value of $t_{m,0}$ increases unboundedly with m. To cope with this issue we decompose the sampling epochs $t_{m,0}$ as follows

$$t_{m,0} = (l_m + \mu_m)T_d ,$$ (3.61)

where

$$l_m = \text{int}\left\{\frac{t_{m,0}}{T_d}\right\} \qquad (3.62)$$

is the integer part of $t_{m,0}$ as measured in *clock ticks* of the sampling frequency f_d, and

$$\mu_m = \text{frac}\left\{\frac{t_{m,0}}{T_d}\right\} \qquad (3.63)$$

is the fractional part of $t_{m,0}$, expressed again in sampling clock intervals. The sampling control signal τ_m provided by the CCTU is updated every symbol interval, and so will also be the two values of l_m and μ_m. By substituting (3.61) into (3.60) we obtain

$$\left(l_{m+1} + \mu_{m+1}\right)T_d = \left(l_m + \mu_m\right)T_d + T_s - \gamma e_m, \qquad (3.64)$$

$$\left(l_{m+1} + \mu_{m+1}\right) = \left(l_m + \mu_m\right) + \frac{T_s}{T_d} - \frac{\gamma}{T_d}e_m = \left(l_m + \mu_m\right) + \eta - \gamma' e_m \qquad (3.65)$$

where $\gamma' = \gamma/T_d$ and $\eta = T_s/T_d$. Taking the integer part of both sides in (3.65) we obtain

$$l_{m+1} = l_m + \eta + \text{int}\left\{\mu_m - \gamma' e_m\right\}, \qquad (3.66)$$

where we have assumed that the oversampling ratio η is an integer value. By taking the fractional part of (3.65) we obtain instead

$$\mu_{m+1} = \text{frac}\left\{\mu_m - \gamma' e_m\right\}. \qquad (3.67)$$

Equation (3.66) can also be cast into the form

$$l_{m+1} - l_m = \eta + \text{int}\left\{\mu_m - \gamma' e_m\right\}, \qquad (3.68)$$

whose right hand side term represents the number of input samples to be shifted into the interpolator until the next output is computed.

Once l_m and μ_m are computed, the output of a third-order interpolator is

$$y(t_{m,n}) = \sum_{i=-2}^{1} C_i(\mu_m) \, x\left[(l_m - i)T_d + n\frac{T_c}{2}\right] \,, \quad 0 \le n \le 2L-1. \qquad (3.69)$$

The meaning of $t_{m,n}$, l_m and μ_m for the third-order interpolator is illustrated in Figure 3-26.

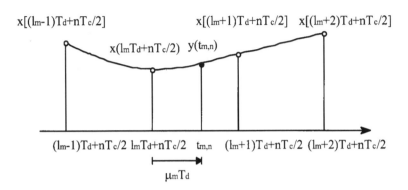

x[(lm-1)Td+nTc/2]   x[(lm+1)Td+nTc/2] x[(lm+2)Td+nTc/2]

x(lmTd+nTc/2) y(tm,n)

(lm-1)Td+nTc/2 lmTd+nTc/2 tm,n   (lm+1)Td+nTc/2 (lm+2)Td+nTc/2

μmTd

Figure 3-26. Meaning of the 3rd-order interpolator parameters.

The coefficients $C_i(\mu)$ ($-2 \le i \le 1$) are given by

$$C_{-2}(\mu) = \frac{\mu^3}{6} - \frac{\mu}{6}, \qquad\qquad\qquad\qquad\qquad (3.70)$$

$$C_{-1}(\mu) = -\frac{\mu^3}{2} + \frac{\mu^2}{2} + \mu, \qquad\qquad\qquad\qquad (3.71)$$

$$C_0(\mu) = \frac{\mu^3}{2} + \mu^2 - \frac{\mu}{2} + 1, \qquad\qquad\qquad\qquad (3.72)$$

$$C_1(\mu) = -\frac{\mu^3}{6} + \frac{\mu^2}{2} - \frac{\mu}{3}, \qquad\qquad\qquad\qquad (3.73)$$

and the block diagram of the interpolator is depicted in Figure 3-27. It is seen that the implementation is that of an FIR filter with variable coefficients.

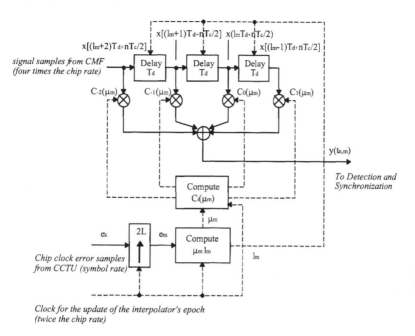

Figure 3-27. Architecture of the 3rd-order interpolator.

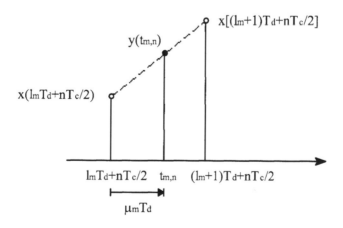

Figure 3-28. Linear interpolation.

A simpler hardware architecture is obtained by resorting to a first-order (linear) interpolator. Interpolation for this simpler case is depicted in Figure 3-28, and the output samples are then given by

$$y\left(t_{m,n}\right) = x\left(l_m T_d + n\frac{T_c}{2}\right)$$

$$+ \mu_m \left\{ x\left[\left(l_m + 1\right)T_d + n\frac{T_c}{2}\right] - x\left(l_m T_d + n\frac{T_c}{2}\right) \right\}. \qquad (3.73)$$

By re-arranging (3.73) we obtain

$$y\left(t_{m,n}\right) = \left(1 - \mu_m\right) x\left(l_m T_d + n\frac{T_c}{2}\right) + \mu_m\, x\left[\left(l_m + 1\right)T_d + n\frac{T_c}{2}\right], \qquad (3.74)$$

which can be cast into a form similar to (3.69)

$$y\left(t_{m,n}\right) = \sum_{i=-1}^{0} C_i\left(\mu_m\right) x\left[\left(l_m - i\right)T_d + n\frac{T_c}{2}\right] \qquad (3.75)$$

with

$$C_0\left(\mu\right) = 1 - \mu, \qquad (3.76)$$

$$C_{-1}\left(\mu\right) = \mu. \qquad (3.77)$$

The architecture of the linear interpolator is finally depicted in Figure 3-29.

When implemented in fixed point arithmetic, the linear interpolators will be affected by quantization errors. The input signal $x(kT_d)$ is replaced by a quantized signal

$$x'\left(kT_d\right) = \left(-1\right)^s \cdot \Delta_x \cdot \sum_{k=0}^{n_C-2} b_k^{(x)} \cdot 2^k, \qquad (3.78)$$

where n is the word length, s denotes the sign bit, $\{b_k^{(x)}\}$, with $0 \le k \le n_C - 2$ and $b_k^{(x)} \in \{0,1\}$, is the code word for the absolute value of $x'(kT_d)$ and Δ_x is the quantization step. The parameters n_C and Δ_x must be chosen so as not to introduce significant distortion in the representation of the samples $x(kT_d)$. In particular, the peak value $x_{\max}$ of the signal should be such that

$$2^{n_C-1}\Delta_x = x_{\max}. \qquad (3.79)$$

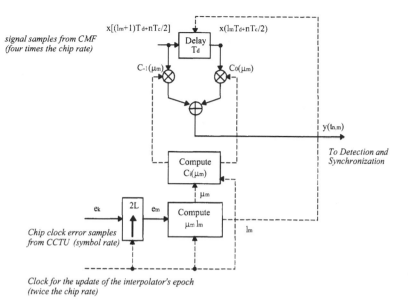

signal samples from CMF
(four times the chip rate)

$x[(lm+1)T_{d}+nT_c/2]$

Delay T_d

$x(lmT_d+nT_c/2)$

$C_{-1}(\mu_m)$ $C_0(\mu_m)$

$y(t_{n,m})$

To Detection and
Synchronization

Compute $C_i(\mu_m)$

μ_m

e_k 2L e_m

Compute $\mu_m\ l_m$

l_m

Chip clock error samples
from CCTU (symbol rate)

Clock for the update of the interpolator's epoch
(twice the chip rate)

Figure 3-29. Architecture of the linear interpolator.

All of the coefficients $C_i(\mu)$ of the interpolator have absolute values smaller than unity for any μ in the interval $[0,1]$. Therefore they can be quantized by n_C bits (as the input samples)

$$C_i'(\mu)=(-1)^s \cdot \Delta_C \cdot \sum_{k=0}^{n_C-2} b_k^{(C_i)} \cdot 2^k \qquad (3.80)$$

with the same notation as above, and where

$$2^{n_C-1}\Delta_C =1. \qquad (3.81)$$

The samples $y(t_n)$ are generated according to (3.54), and their quantized version $y'(t_n)$ is then

$$y'(t_n)=(-1)^s \cdot \Delta_C \cdot \Delta_x \cdot \sum_{k=0}^{2n_C-1} b_k^{(y)} \cdot 2^k . \qquad (3.82)$$

Our results indicate that the quantization step in the representation of $y'(t_n)$ can still be assumed to be equal to Δ_x with no significant performance impairment. This suggests that some bits can be dropped in the expression of $y'(t_n)$ in order to obtain a signal representation with the same complexity as for $x'(kT_d)$. As is seen in Figure 3-30, the $n_C -1$ LSBs and the two MSBs are neglected. Word length reduction is carried out so as to emulate signal clipping as follows:

i)   if both MSBs are zero, then the remaining $n_C - 1$ bits are left unchanged;

ii) if at least one MSB is nonzero, then all of the remaining $n_C - 1$ bits are set to 1.

Once the $n_C - 1$ LSBs and the two MSBs are dropped, the real and the imaginary parts of $y(t_n)$ can be represented as follows

$$ y'(t_n) = (-1)^s \cdot \Delta_x \cdot \sum_{k=0}^{n_C-2} b_k^{(y)} \cdot 2^k . \tag{3.83} $$

Clearly the binary symbols $b_k^{(y)}$ in (3.82) and (3.83) are not the same, but we preferred to retain the same notation for simplicity.

As discussed above the outputs of the I and Q interpolators, running at the rate $2R_c$, are eventually demultiplexed in two low rate (R_c) signals by two demultiplexers, yielding the prompt and E/L sample streams. The outputs of the interpolators at twice the chip rate ($2R_c$) are also used by the CCAU (Section 2.1.1) for coarse code timing recovery. In this code acquisition mode the CCTU is inactive, and the sampling epoch of the interpolators are arbitrary and constant in time.

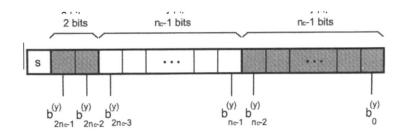

Figure 3-30. Bit reduction in the representation of the samples at the interpolator output.

2.3   Carrier Synchronization

2.3.1   Carrier Frequency Synchronization

Initial code acquisition may be severely impaired when the initial carrier frequency offset, denoted as v, owed to residual Doppler shift and/or instability of the local oscillator in the receiver is comparable to the inverse of the symbol period T_s. Frequency offset causes a sort of 'decorrelation' of the observed signal within the coherent integration window of the serial acquisition scheme discussed above that can be quantified in a power loss figure. For instance, it can be shown [Syn98] that a carrier frequency offset equal to half the symbol rate yields a coherent integration loss of about 4 dB, and far higher losses have to be taken into account for larger frequency errors. Unfortunately, estimation of the carrier frequency offset cannot be carried out reliably unless code is coarsely acquired. This 'chicken or egg' problem has no simple solution: the only viable approach is a sort of joint two-dimensional time/frequency acquisition over the possible code epochs and over a number of pre-determined frequency bins within an assigned uncertainty interval.

Initial frequency uncertainty is especially an issue when dealing with Low or Medium Earth Orbiting satellites (LEO/MEO). Even feeder link pre-compensation techniques will not prevent the residual downlink Doppler shift from being larger than the symbol rate in coded voice communications. The difficulty of carrier frequency acquisition is another facet of the wideband characteristic of the DS/SS signal. Actually, both in narrowband and in SS modulations one has to determine the carrier frequency with an accuracy much smaller than the symbol rate to ensure good data detection. Clearly this estimation task is apparently harder to accomplish when the bandwidth of the observed signal is many times greater than the symbol rate, as in wideband modulation. A survey of synchronization techniques for CDMA transmissions is presented in [Syn98].

Upon completion of raw acquisition of initial code phase and carrier frequency, the (small) residual offset can be removed at baseband on the symbol rate signal resulting form despreading/accumulation. The raw frequency offset estimated during the acquisition phase as above is used to correct the local oscillator frequency, and the residual frequency error after despreading is reduced to a small fraction of the symbol rate. Fine frequency offset compensation can be performed resorting to conventional closed loop frequency estimation/correction techniques [DAn94]. In particular, a few algorithms have been analyzed and experimented with in CDMA modems:

i) Rotational Frequency Detector (RFD) [Cla93];
ii) Dual Filter Discriminator (DFD) [Alb89];
iii) Cross-Product AFC (CP-AFC) [Nat89];
iv) Angle Doubling AFC (AD-AFC).

Further details on frequency error detectors to be implemented in a digital receiver are found in the overview paper [Moe94] and a further example of such techniques is also described in [DAn94].

Figure 3-31 shows the overall architecture of the AFCU implemented in the MUSIC receiver, with an indication about the different processing rates in the different circuit parts: f_c represents here the sampling rate of the ADC, f_s is the symbol rate, L is as usual the spreading factor, and $N = 4\rho$ is the number of samples per chip.

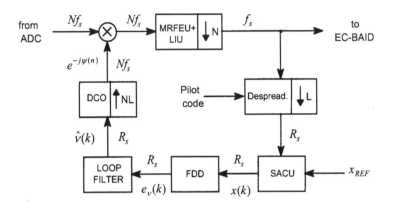

Figure 3-31. Block diagram of the AFC unit.

As is apparent, we use here a 'long loop' approach, in which the Multi-Rate Front End Unit (MRFEU) is encompassed by the loop. This does not harm loop stability, since the relevant processing latency is definitely negligible with respect to the intrinsic response time of the AFCU as a whole. The output on time samples of the LIU at chip rate are despread with the pilot code. The resulting symbol time samples are processed by the SACU and are subsequently sent to the Frequency Difference Detector (FDD). The latter outputs in turn the frequency error signal $e_v(k)$ which is filtered by the loop filter according to the following recursive equation, to give the updated estimate $\hat{v}(k)$ (at symbol rate[1]) of the frequency offset v

$$\hat{v}(k+1) = \hat{v}(k) - \gamma e_v(k),\qquad\qquad(3.84)$$

where γ is the step size of the frequency tracking algorithm (which in the following will be also referred as γ_{AFCU}). The frequency estimate $\hat{v}(k)$

---

[1] When using a pilot channel to perform frequency control, we could also lengthen our coherent despreading interval with respect to a symbol period, and slow down accordingly the updating rate. This would probably make the loop more robust to noise, but makes it more sensitive to a large initial frequency offset, which can be in the MUSIC receiver as high as 10% of the symbol rate. This is why here we have stuck to symbol time integration and symbol rate updating.

drives the DCO which in turn provides the complex-valued oscillation $\exp\{-j\psi(n)\}$ at the ADC sampling rate. The instantaneous phase $\psi(n)$ of such oscillation is generated by the DCO according to the recursive equation (running of course at sampling frequency rate)

$$\psi(n+1)=\psi(n)+\frac{2\pi\hat{v}(k)}{Nf_c}\ \ \text{mod}(2\pi), \tag{3.85}$$

where the sampling-rate index n is related to the symbol rate index k according to $k=\text{int}(n/NL)$. The counter-rotating complex-valued exponential generated by the DCO with instantaneous phase (3.85) is used by the digital downconverter in Figure 3-7 to remove the frequency offset from the I/Q received samples. If we denote with $x(k)$ the symbol time signal at the SACU output (see Figure 3-31), the frequency error signal $e_v(k)$ provided by the FDD (see Figure 3-32) is

$$e_v(k)=\Im\left\{x(k)x^*(k-2)\right\}. \tag{3.86}$$

Assuming ideal chip timing recovery and neglecting for the moment the contribution of channel noise, we have

$$x(k)=Ae^{j(2\pi vkT_s+\varphi)}+\eta(k) \tag{3.87}$$

where v is the frequency offset, T_s is the symbol interval and $\eta(k)$ is the MAI contribution. With (3.87) in (3.86) we find

$$e_v(k)=A^2\sin(4\pi vT_s)+\mu(k), \tag{3.88}$$

where

$$\mu(k)=\eta(k)e^{-j[2\pi v(k-2)T_s+\varphi]}+\eta^*(k-2)e^{j(2\pi vkT_s+\varphi)}$$
$$+\eta(k)\eta^*(k-2) \tag{3.89}$$

We want now to obtain an expression for the average characteristics of the FDD (the so called *S curve*). Therefore we compute the statistical expectation of $e_v(k)$ conditioned on a fixed value of the frequency offset v. This computation is easily done if we observe that $\eta(k)$ and $\eta(k-2)$ have zero mean and are *statistically independent*.

The latter property comes from the observation that two MAI samples taken with a two-symbol delay are necessarily independent, since the correlation time of a DS/SS signal is roughly equal to one symbol interval (neglecting the tails of the chip shaping pulse) owing to data modulation. In summary, the S curve of the FDD (3.88) is

$$S(v) = E\{e_v(k)|v\} = A^2 \sin(4\pi v T_s).$$ (3.90)

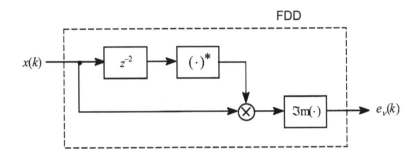

Figure 3-32. Block diagram of the FDD.

The 'lock in' (also termed 'pull in') range of the AFCU is thus equal to $0.25/T_s$. The reason why we *did not* resort to the customary 'delay and multiply' FDD should be now clear. Such FDD reads, in fact,

$$e_v(k) = \Im\{x(k)x^*(k-1)\},$$ (3.91)

and, as can be easily shown following the same reasoning as above, has an operating range twice as large as FDD (3.88). Unfortunately the MAI terms $\eta(k)$ and $\eta(k-1)$ appearing in (3.89) would be in general *correlated*, and thus the term $\mu(k)$ in the expression of the S curve would have *non-zero* mean. This would give rise to a *bias* term with detrimental effect on the AFCU performance.

Figure 3-33 shows a sample of a typical AFCU acquisition transient with ideal chip timing recovery, and obtained in the following conditions:

i) $L = 64$;
ii) $K = 32$ active users;
iii) $E_b/N_0 = 2$ dB;
iv) $C/I = -6$ dB;
v) $P/C = 6$ dB;
vi) $vT_s = 0.1$;

vii) $\gamma_{AFCU} = 2^{-15} - 2^{-19}$

Notice that the step size is set to $\gamma_{AFCU} = 2^{-15}$ during an initial 10^4 symbol interval, and then it is reduced down to the value 2^{-19} to yield low steady state jitter. Without such reducing of the step size, the steady state jitter would cause cycle slips in the Carrier Phase Recovery Unit (CPRU). As is seen, the acquisition time is roughly equal to 10^4 symbol intervals. We also show in Figure 3-34 a curve of the steady state RMS frequency estimation error normalized to the symbol rate as a function of the E_b / N_0 ratio and with the same parameters as in Figure 3-33.

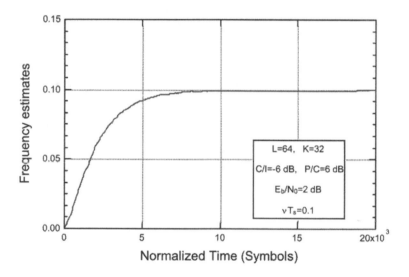

Figure 3-33. AFCU acquisition sample.

2.3.2   Carrier Phase Synchronization

A *coherent* receiver for DS/SS CDMA signals has to perform carrier phase estimation and correction. Broadly speaking, phase correction could be implemented at IF, with an NCO driven by an appropriate error signal derived from the baseband signal components.

However, the use of a non-phase locked conversion oscillator may be favored both to possibly save on the NCO and to solve possible stability issues in the use of a 'long loop' approach. Phase recovery can be thus implemented digitally on the baseband components of the received signal either using a symbol rate closed loop tracker or an open loop estimator [Men97]. We will not dwell further on the different algorithms for phase error detection to be employed in the CDMA receiver, since they do not bear any peculiar aspect with respect to standard techniques for narrowband linear modulations [Men97].

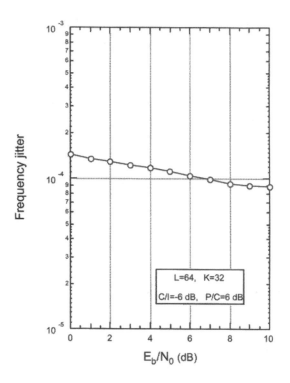

Figure 3-34. Normalized RMS jitter of the AFCU frequency estimate (symbol rate units).

In the design of the MUSIC receiver we have resorted to a conventional decision directed phase recovery unit (labeled CPRU) for coherent data demodulation at the output of the EC-BAID. We had here two alternative architectures: on the one hand, we could perform efficient data aided phase recovery on the pilot channel (trusting it stays coherent with the useful traffic channel) with no protection whatsoever against possible MAI; on the other hand, we could resort to 'blind' (i.e., non-data aided) decision directed recovery on the traffic channel *on the signal at the output of the EC-BAID*, thus in a condition of good immunity from MAI. We argued that the latter solution is preferable. In fact, the advantage of a data aided vs. a decision directed loop vanishes when the SNIR on the observed signal is good. This is the case of the EC-BAID output, but is *not* the case of the un-protected pilot channel that can rely on the pilot power margin P/C, but has no immunity to MAI.

Figure 3-35 shows the overall architecture of the CPRU [Men97], with components all operating at symbol rate R_s. Signal $y(k)$ coming out of EC-

BAID is first counter-rotated by an amount equal to the current estimate $\hat{\theta}(k)$ of the carrier phase θ as follows

$$z(k) = y(k)e^{-j\hat{\theta}(k)} \tag{3.92}$$

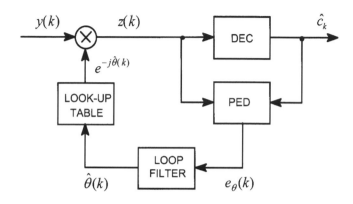

Figure 3-35. Block diagram of the CPRU.

Subsequently it is sent to a hard detector so as to provide an estimate $\hat{c}_k$ of the k-th QPSK data symbol. Signals $z(k)$ and $\hat{c}_k$ are used by the Phase Error Detector (PED) to build up the phase error signal

$$e_\theta(k) = -\Im\{\hat{c}_k^* z(k)\} \tag{3.93}$$

which represents the input of the loop filter. We resorted to a second-order loop in order (to compensate for the residual frequency jitter at the output of the AFCU) whose loop equations are

$$\hat{\theta}(k+1) = \hat{\theta}(k) + \mu(k), \tag{3.94}$$

$$\mu(k) = \mu(k-1) - \gamma(1+\rho)e_\theta(k) + \gamma e_\theta(k-1), \tag{3.95}$$

as depicted in Figure 3-36.

The two loop parameters γ and ρ (which in the following will be also referred as γ_{CPRU} and ρ_{CPRU}, respectively) can be related to the noise loop bandwidth B_L and to the loop damping factor ς. When $B_L T_s \ll 1$ (as is always the case) one can resort to the following approximate equations

$$\rho \approx \frac{4B_L T_s}{1+4\varsigma^2}, \qquad \gamma \approx \frac{16\varsigma^2 B_L T_s}{A\left(1+4\varsigma^2\right)}, \qquad (3.96)$$

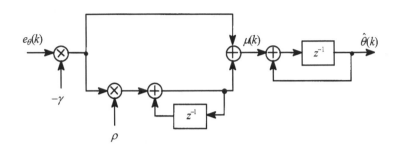

Figure 3-36. Second-order loop.

where A is the slope of the PED average characteristics (*S curve*) at the loop equilibrium point. The phase estimate (3.94) is input to a LUT similar to the one in Figure 3-8 to provide the counter-rotation factor $\exp\{-j\hat{\theta}(k)\}$ that compensates for the phase error on $y(k)$.

As with all decision directed loops, the QPSK hard detector in Figure 3-35 (labeled 'DEC') introduces a $\pi/2$ *ambiguity* in the phase recovery process. This means that the two I/Q components may be swapped and/or they can be in reversed sign after phase correction. This ambiguity is solved with the aid of a 24 bit *Unique Word* (UW) which is periodically inserted in the data frame to ease frame synchronization.

Figure 3-37 shows a typical sample phase acquisition with the PLL as in Figure 3-35 operating in tandem with the EC-BAID, assuming ideal frequency/timing references, and with the following parameters:

i) $L = 64$;
ii) $K = 32$ active users;
iii) $E_b / N_0 = 2$ dB;
iv) $C / I = -6$ dB;
v) $P / C = 6$ dB;
vi) $\gamma_{CPRU} = \rho_{CPRU} = 2^{-9}$. This roughly corresponds (considering the average signal amplitude at the output of the EC-BAID in the steady state) to a normalized loop bandwidth $B_L T_s = 10^{-3}$ and a damping factor $\varsigma = 0.8$;
vii) EC-BAID: $\gamma_{BAID} = 2^{-15}$;
viii) initial phase error $\Delta\theta = 30$ deg.

The acquisition time is roughly equal to 1.5×10^4 symbol intervals. Admittedly, this time is somewhat larger than expected, considering the damping factor and loop bandwidth as above. On the other hand, what we see in Figure 3-37 is actually a *joint* acquisition of the CPRU and the EC-BAID. During the first stage of acquisition, the signal amplitude at the EC-BAID output is *smaller* than at steady state, so that the CPRU is pretty slower than predicted considering steady state values.

Figure 3-38 shows the RMS residual steady state phase jitter at the CPRU output as a function of the E_b / N_0 ratio with the same parameter values as in Figure 3-37.

As an overall summary of carrier recovery subsystems performance, we show in Figure 3-39 some simulation results concerning the BER of the EC-BAID receiver equipped with AFCU (steady state configuration, $\gamma_{AFCU} = 2^{-19}$) and CPRU as described above. Such results (dots interpolated by lines) are compared to what is obtained assuming ideal frequency/phase recovery. The degradation with respect to the ideal case is immaterial. We could not also detect any cycle slip event in our simulations with the selected AFCU/CPRU configuration. Our simulation trials had a maximum length of 10^7 symbols. We could detect a phase lock loss event only with $E_b / N_0 = 0$ dB *and* after intentionally increasing the steady state step size to $\gamma_{AFCU} = 2^{-18}$.

In the implementation of the MUSIC receiver (as in any other data demodulator) a fundamental function is detection of the phase lock condition to validate the demodulated data with respect to the upper network layers. Therefore we have also focused on the design of a reliable phase lock detector to perform such functions. The detector has to be *blind* to operate on the EC-BAID output and sufficiently robust not to incur in false alarm or missed lock events. To this aim we have built up a 'lock metrics' $l(k)$ as follows

$$l(k+1) = (1 - \gamma_L) l(k) + \gamma_L \left[|z_I(k)| + |z_Q(k)| \right]. \tag{3.97}$$

Here $z_I(k)$ and $z_Q(k)$ are the real and imaginary parts of the signal $z(k)$ in Figure 3-35, respectively, and γ_L is a *forgetting factor*. In practice $l(k)$ is a low pass filtered version of the signal $\tilde{z}(k) = |z_I(k)| + |z_Q(k)|$, the forgetting factor γ_L determining the filter bandwidth. Equation (3.97) was designed bearing in mind a low complexity constraint: $z(k)$ is already available in the receiver, and absolute value extraction is trivial on the digital representation of the relevant I/Q components. The rationale behind (3.97) is easily explained. If we neglect noise and MAI we have (see Figure 3-35)

$$z(k) = A c_k e^{j\Delta\theta(k)}, \tag{3.98}$$

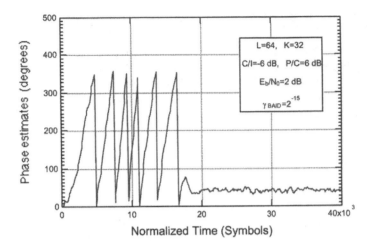

Figure 3-37. CPRU acquisition sample.

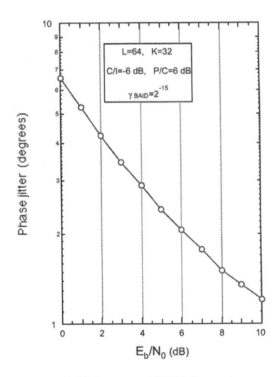

Figure 3-38. Accuracy of CPRU phase estimates.

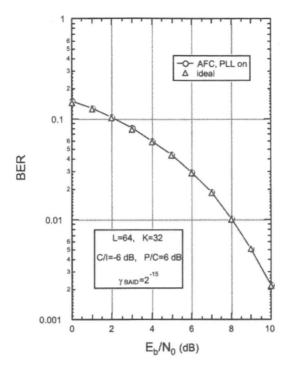

Figure 3-39. BER performance in the presence of frequency and phase errors.

where $\Delta\theta(k) = \theta(k) - \hat{\theta}(k)$ is the residual phase error at step k, and $c_k \in \{\pm 1 \pm j\}$ is the kth transmitted QPSK symbol on the useful traffic channel. If we look at $\tilde{z}(k)$ as a function of $\Delta\theta(k)$ we easily find that it is *not* dependent on the particular value of c_k, and it is periodic with period $\pi/2$

$$\tilde{z}(k) = A\{\left|\cos[\Delta\theta(k)] + \sin[\Delta\theta(k)]\right| + \left|\cos[\Delta\theta(k)] - \sin[\Delta\theta(k)]\right|\}$$

(3.99)

(recall that $A > 0$ by definition). As is seen from the plot of (3.99) in Figure 3-40, $\tilde{z}(k)$ attains its maximum value $2A$ when the phase error is a multiple of $\pi/2$, i.e., when the phase loop is in lock. Re-considering noise and MAI, $\tilde{z}(k)$ needs filtering to yield a reliable lock metrics as in (3.97).

Before the AFCU and the CPRU have attained lock $\tilde{z}(k)$ is affected by a frequency offset. In such a condition $\Delta\theta(k)$ has a linear evolution with time, and therefore the oscillating plot in Figure 3.40 is in a sense 'swept' on the phase x-axis. If the forgetting factor is small, i.e., $\gamma_L \ll 1$, the lock metrics

$l(k)$ in (3.97) tends to be equal to the time averaged value of $\tilde{z}(k)$

$$l(k) \approx \frac{1}{\pi/2} \int_{-\pi/4}^{\pi/4} 2A\cos(\Delta\theta)\, d\Delta\theta \approx 1.8A. \tag{3.100}$$

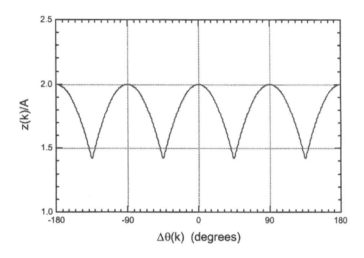

Figure 3-40. $\tilde{z}(k)$ vs. the phase error $\Delta\theta(k)$.

Our lock detection criterion will be a comparison of $l(k)$ with a suited threshold ranging between $1.8A$ and $2A$. If the threshold is crossed, the phase error should be stable and close to one of the four lock point multiples of $\pi/2$.

Figures 3-41 and 3-42 show the evolution of the lock metrics and of the AFCU frequency estimate starting from receiver switch on in the following condition:

i) $L = 64$;
ii) $K = 32$ active users;
iii) $E_b / N_0 = 2$ dB;
iv) $C / I = -6$ dB;
v) $P / C = 6$ dB;
vi)   AFCU: $\gamma_{AFCU} = 2^{-19}$, $vT_s = 0.1$. The frequency step size is intentionally set from the very start to its steady state value. This has the effect of lengthening the frequency acquisition time to show better the two different DC levels attained by $l(k)$ in the two different out of lock and in lock conditions;
vii) CPRU: $\gamma_{CPRU} = \rho_{CPRU} = 2^{-9}$;

viii) EC-BAID: $\gamma_{BAID} = 2^{-15}$;
ix) lock detector: $\gamma_L = 2^{-13}$.

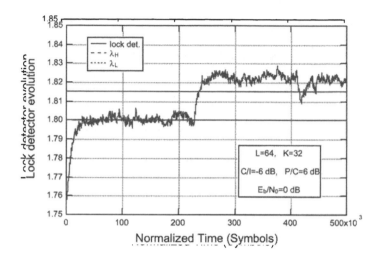

Figure 3-41. Lock metrics evolution @ $E_b / N_0 = 0$ dB.

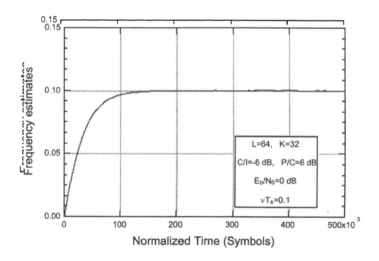

Figure 3-42. Frequency acquisition @ $E_b / N_0 = 0$ dB.

Joint evaluation of Figures 3-41 and 3-42 is quite instructive. It is seen that in a first stage the frequency error is quite large ($vT_s = -0.10$), the CPRU has no way to lock in, and the lock metrics (initialized at $l(0) = 1.75$) have a short acquisition and settles at the expected out of lock value 1.8. As soon as the AFCU acquisition is over, and thus the frequency error is small (roughly $k = 2 \cdot 10^5$), the CPRU starts acquiring lock, and in parallel (after a short CPRU acquisition time) the lock metrics rapidly attains the lock value 1.82. Unfortunately this value is substantially smaller than the theoretical peak value of 2 in Figure 3-40 owed to noise induced *biasing*. We can therefore use a strategy of comparison with hysteresis to detect "out of lock→in lock" and "in lock→out of lock" transitions based on the two threshold values $\lambda_L = 1.8$ and $\lambda_H = 1.815$. This prevents the circuit to detect false events like the one we would find in Figure 3-43 at $k \cong 4.2 \times 10^5$ should we use a single threshold at λ_H with no hysteresis.

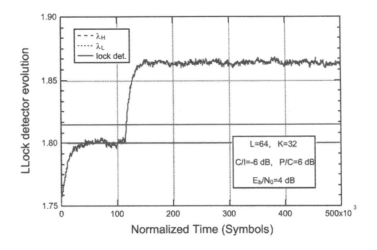

Figure 3-43. Lock metrics evolution @ $E_b / N_0 = 0$ dB.

Concerning the *bias* phenomenon for the out of lock and in lock values of $l(k)$ mentioned above, we found that the out of lock value 1.8 is very marginally affected by the operating condition in terms of SNIR, probably owing to the implicit time averaging effect on $\tilde{z}(k)$ we have discussed. Instead, the in lock value tends to grow when the SNIR improves. Thus, the same threshold values determined for the worst case in Figure 3-43 can be safely re-used in conditions of better SNIR.

3.   SIGNAL DETECTION AND INTERFERENCE MITIGATION

Implementation of a single-channel interference mitigating CDMA detector represents the main novelty of the MUSIC project. In this Section we present the interference mitigating feature of the MUSIC receiver which is based on the EC-BAID algorithm to be detailed hereafter.

3.1 EC-BAID Architecture

We start with the analytical description of the signal at the receiver input, assuming that K user traffic channels in DS/SS format are code multiplexed in A-CDMA mode (see Chapter 2). The generic kth CDMA user transmits a stream of complex-valued information bearing symbols, denoted as $\tilde{a}_k(u) = a_{k,p}(u) + ja_{k,q}(u)$. The symbols, which belong to a QPSK alphabet (i.e., $a_{k,p}(u), a_{k,q}(u) \in \{\pm 1\}$) and run at symbol rate $R_s = 1/T_s$, are spread over the frequency spectrum by multiplication with a binary signature code, denoted as $c_k(\ell) \in \{\pm 1\}$, with period L and running at chip rate $R_c = 1/T_c$. The signature is actually a short code as its repetition period L spans exactly one symbol interval: $T_s = L \cdot T_c$. Chip rate symbols are eventually shaped by a transmit filter with SRRC impulse response $g_T(t)$. At the receiver side, after baseband conversion, the overall signal, denoted as $r(t)$, is made of K CDMA channels plus additive noise $n(t)$ as follows

$$r(t) = \sum_{k=1}^{K} \sum_{u=-\infty}^{\infty} \sqrt{P_k}\,\tilde{a}_k(u) s_k\left(mT_c - uT_s - \tau_k\right)$$
$$\cdot \exp\{j(2\pi v_k mT_c + \phi_k)\} + n(t), \tag{3.101}$$

where P_k is the RF power of the kth channel and $s_k(t)$ is the relevant spreading signature defined as

$$s_k(t) = \sum_{\ell=0}^{L-1} c_k(\ell) g_T\left(t - \ell T_c\right). \tag{3.102}$$

In (3.101) τ_k, ϕ_k and v_k are the time delay, the carrier phase shift, and the frequency offset of the generic k-th traffic channel w.r.t. the useful traffic signal, which, without loss of generality is assumed to be channel 1. We assume for now that the carrier frequency error relevant to channel 1 is perfectly compensated for by means of an ideal AFC subsystem (i.e., $\Delta f_1 = 0$) and that perfect chip timing recovery is performed (i.e., $\tau_1 = 0$). The signal $r(t)$ is then sent through a baseband filter with impulse response $g_R(t)$ performing Nyquist's SRRC chip matched filtering, followed by chip time sampling (or interpolation in the case of a digital implementation). The signal samples taken at time $t_m = m \cdot T_c$ at the output of the CMF are thus

$$y(m) = r(t) \otimes g_R(t)|_{t=mT_c}. \tag{3.103}$$

The chip time signal $y(m)$ is then input to the EC-BAID data detector that was introduced in Section 2-5. We will described here the detector in more detail, starting back from the very fundamentals, just to make this section as much self-contained as possible. As detailed in [Rom97], the EC-BAID uses a three-symbol observation window to detect one information bearing symbol. In the subsequent analytical description we will use the superscript ^{e} to denote a $3L$-dimensional vector (also termed 'extended vector' as opposed to 'non-extended' L-dimensional vectors), the superscript ^{T} to denote transposition, and the asterisk * to denote complex conjugation. The $3L$-dimensional array of CMF samples observed by the detector is given by

$$\mathbf{y}^e(r) = \begin{bmatrix} y_0^e \\ \vdots \\ y_{3L-1}^e \end{bmatrix} = \begin{bmatrix} \mathbf{y}_{-1}(r) \\ \mathbf{y}_0(r) \\ \mathbf{y}_1(r) \end{bmatrix}, \tag{3.104}$$

where

$$\mathbf{y}_w(r) = \begin{bmatrix} y[(r+w)LT_c] \\ y[((r+w)L+1)T_c] \\ y[((r+w)L+2)T_c] \\ \cdots\cdots\cdots \\ y[((r+w)L+L-1)T_c] \end{bmatrix} \tag{3.105}$$

with $w = -1,0,1$. The EC-BAID is a linear detector operating on the chip rate sampled received signal $y(m)$ to yield the symbol rate signal $b(r)$ as follows

$$b(r) = \frac{1}{L}\mathbf{h}^e(r)^T \mathbf{y}^e(r), \tag{3.106}$$

where $\mathbf{h}^e(r)$ is the $3L$-dimensional array of the *complex-valued* detector coefficients. It is apparent that detection of each symbol calls for observation of *three* symbol periods (i.e., the current, the leading, and the trailing ones) which represent the so called *observation window* (W_{LEN}). This suggests the three-fold parallel implementation of the detector sketched in Figure 2-20, and repetead here in Figure 3-44, wherein the first detector unit processes the $(r-1)$th, the rth and the $(r+1)$th symbol periods for the detection of the rth symbol, the second unit processes the rth, the $(r+1)$th and the $(r+2)$th periods, for the detection of the $(r+1)$th symbol, and the third unit processes the $(r+1)$th, the $(r+2)$th and the $(r+3)$th periods, for the detection of the $(r+2)$th symbol. The structure of the detector units will be outlined in the sequel. Also, in the algorithm description we will assume a normalized observation window $W_{LEN} = 3$, whilst further considerations about the selection of the optimum value of W_{LEN} will be reported later in Section 4.1.

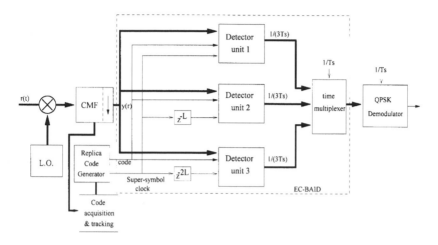

Figure 3-44. EC-BAID top level functional block.

The output stream of soft values for data detection is obtained by sequentially selecting the three detector unit outputs at rate $1/T_s$ by means of a multiplexer. We need thus a further clock reference ticking at the so called *Super-Symbol* rate $R_{SS} = 1/(3T_s)$, i.e., once every three symbols. Taking this into account, the sample at the output of the n-th detector unit ($n = 1, 2, 3$) is

$$b(3s + n - 1) = \frac{1}{L} \mathbf{h}^{e,n}(s)^T \mathbf{y}^e(3s + n - 1),$$
(3.107)

with s running at super-symbol rate. To achieve blind adaptation the complex coefficients $\mathbf{h}^{e,n}$ of each detector are *anchored* to the user signature sequence, represented by the L–dimensional array $\mathbf{c}$ containing the chips $c_1(\ell)$ of the useful signal 1. The anchoring condition is obtained as follows [Rom97]. First, we decompose $\mathbf{h}^{e,n}$ in two parts

$$\mathbf{h}^{e,n}(s) = \mathbf{c}^e + \mathbf{x}^{e,n}(s),$$
(3.108)

where

$$
\mathbf{c}^e = \begin{bmatrix} \mathbf{0} \\ \mathbf{c} \\ \mathbf{0} \end{bmatrix}, \quad
\mathbf{c} = \begin{bmatrix} c_0 \\ c_1 \\ \vdots \\ c_{L-1} \end{bmatrix}, \quad
\mathbf{x}^{e,n}(s) = \begin{bmatrix} x_0^{e,n} \\ \vdots \\ x_{3L-1}^{e,n} \end{bmatrix} = \begin{bmatrix} \mathbf{x}_{-1}^n(s) \\ \mathbf{x}_0^n(s) \\ \mathbf{x}_1^n(s) \end{bmatrix}. \tag{3.109}
$$

where we set $c_i = c_1(i)$ for simplicity. We impose then the following 'anchor' constraint

$$
\mathbf{c}^T \mathbf{x}_w^n = 0 \tag{3.110}
$$

with $w = -1, 0, 1$, and the optimum MMOE configuration of the detector is found through application of a recursive update rule for the detector coefficients. As is detailed in [Rom97], the error signal in the recursion for detector n is given by

$$
\mathbf{e}^{e,n}(s) = \begin{bmatrix} \mathbf{e}_{-1}^n(s) \\ \mathbf{e}_0^n(s) \\ \mathbf{e}_1^n(s) \end{bmatrix}, \tag{3.111}
$$

where

$$
\mathbf{e}_w^n(s) = b(3s + n - 1)\left[\mathbf{y}_w^*(3s + n - 1) - \frac{\mathbf{y}_w^*(3s + n - 1)^T \mathbf{c}}{L} \mathbf{c} \right] \tag{3.112}
$$

$w = -1, 0, 1$. If the three detector units were running independently, the update equation for each detector would simply be [Rom97]

$$
\mathbf{x}^{e,n}(s+1) = \mathbf{x}^{e,n}(s) - \gamma \mathbf{e}^{e,n}(s), \tag{3.113}
$$

with s ticking at super-symbol rate and where γ is the adaptation step (which in the following will be also referred as γ_{BAID}).

Equation (3.110) forces the so called 'chunk' orthogonality condition on all three adaptive detector components $\mathbf{x}_w^n$, leading to a detector which we call EC-BAID-I, whose structure is outlined in Figure 3-45. On the other hand, we recognize that there is little information about the symbol to be detected in the signal segments spanned by $\mathbf{x}_{-1}^n$ and $\mathbf{x}_1^n$. Therefore we can also limit the orthogonality constraint to $\mathbf{x}_0^n$ only, i.e., $\mathbf{c}^T \mathbf{x}_0^n = 0$. In so doing, the components $\mathbf{x}_{1,-1}^n$ and $\mathbf{x}_{1,1}^n$ have more degrees of freedom for minimizing the selected error cost function as detailed in [Rom97]. Such a modified EC-BAID algorithm, dubbed EC-BAID-II, is formalized by

$$\mathbf{x}^{e,n}\left[3(s+1)\right] = \mathbf{x}^{e,n}(3s) - \gamma b(3s+n-1)$$

$$\cdot\left[\mathbf{y}^{e*}(3s+n-1) - \frac{\mathbf{y}^{e*}(3s+n-1)^{T}\mathbf{c}^{e}}{L}\,\mathbf{c}^{e}\right] \quad (3.114)$$

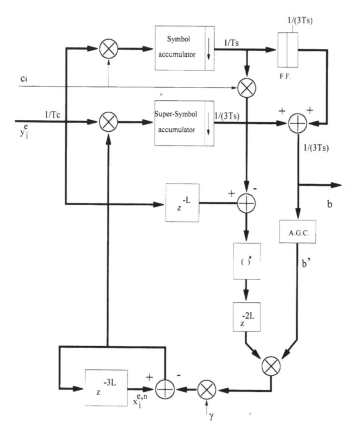

Figure 3-45. EC-BAID-I detector: $x_i^{e,n}$ and y_i^e are the elements of $\mathbf{x}^{e,n}$ and $\mathbf{y}^e$, respectively.

with the following reduced anchoring condition

$$\mathbf{c}^{e^{T}}\mathbf{x}^{e,n} = 0. \quad (3.115)$$

The EC-BAID-II (whose architecture is depicted in Figure 3-46) reveals enhanced robustness against MAI [Rom97]. On the other hand, the EC-BAID-I is more resilient to the lack of randomness for the modulating data and thus it can be conveniently used in all situations where no data scrambling is possible. For this reason the proposed MUSIC receiver architecture allows for EC-BAID type I or II programmability by setting a proper input control signal.As shown above, the EC-BAID-I and -II versions (Figures 3-45 and 3-46) require in principle three separate units, each with its own 'local' copy of $\mathbf{x}^{e,n}$. This is not necessarily true if we work out different variants of the update algorithm (3.113) and of the output computer (3.106).

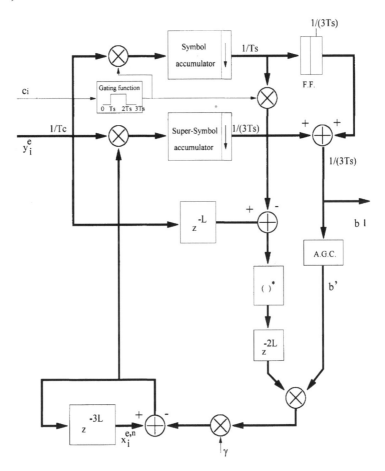

Figure 3-46. EC-BAID-II detector: $x_i^{e,n}$ and y_i^e are the elements of $\mathbf{x}^{e,n}$ and $\mathbf{y}^e$, respectively.

The final architecture of EC-BAID-I and –II, whose top level diagram sketched in Figure 3-47, follows in fact the so called 'Select and Add' (S&A) arrangement. In particular, the S&A architecture uses a clock of period $T_c/3$ (over-clock) to re-use the arithmetical part of the circuit three times for each symbol period. This allows to calculate the output $b(r)_s$ get the entire dot product (3.106) and update, a single, 'unique' vector $\mathbf{x}^e$, shared by all of the three pipelined detectors, and all this in a single period T_s.

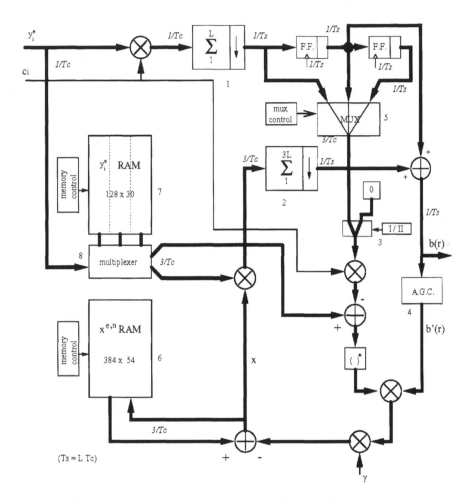

Figure 3-47. Optimized architecture of the EC-BAID 'Select and Add'.

As is depicted in Figure 3-47, blocks 1 and 2 evaluate the correlations $\mathbf{y}(r)^T \mathbf{c}_1$ and $\mathbf{x}^e(r)^T \mathbf{y}^e(r)$, respectively, yielding the output strobe $b(r)$ at symbol rate. The vector $\mathbf{x}^e$ is stored in memory (item 6 in Figure 3-47) and each of its *3L* elements is updated every $T_c/3$. In particular, during the *i*-th chip period within the *r*-th symbol interval, the coefficients of $\mathbf{x}^e$ relevant to the *i*-th chip of $\mathbf{y}(r-1)$, $\mathbf{y}(r)$, and $\mathbf{y}(r+1)$ are updated. A dedicated memory (item 7) is used to store the most recent *3L* input chips. Multiplexers 5 and 8 properly re-align the internal dataflow, while multiplexer 3 selects the desired EC-BAID algorithm version (type-I or -II). The update equation for EC-BAID-II is

$$\mathbf{x}_w(r+1) = \mathbf{x}_w(r) - \gamma b(r-1) \left[\mathbf{y}_w^*(r-1) - \frac{\mathbf{y}_w^*(r-1)^T \mathbf{c}_1}{L} \; \mathbf{c}_1 \delta_w \right], \quad (3.116)$$

where $w = -1, 0, 1$ and δ_w is a Kronecker delta such that $\delta_w = 1$ if $w = 0$ and $\delta_w = 0$ if $w = \pm 1$. Therefore the output of multiplexer 3 is set to zero (instead of to $\mathbf{y}_w^T \mathbf{c}$) for EC-BAID-II (see Figure 3-47). The timing diagram of the S&A main signals is shown in Figure 3-48. The Automatic Gain Control (AGC) on the feedback loop (block 4) is needed in order to keep the amplitude $|b|$ of the output signal constant, irrespective of the different SNIR operating condition.

The equations relevant to the S&A implementation of the EC-BAID-I and –II are summarized in Table 3-3, where r is a symbol time index.

3.2   EC-BAID Optimization

The fixed point ASIC implementation of the S&A introduces truncation errors with respect to theoretical performance which is computed assuming floating point arithmetic. Our adaptive architecture is based on a feedback loop, and so the quantization errors may have detrimental effects on the overall algorithm convergence. In particular, quantization effects may destroy the orthogonality between the vector of the error signals $\mathbf{e}^e$, which is used to generate the adaptive vector $\mathbf{x}^e$, and the code sequence vector $\mathbf{c}^e$. As a consequence, owing to finite arithmetic, the components of $\mathbf{x}^e$ may drift and indefinitely increase, thus causing in the long run saturation and failure of the detector. To prevent this, it is mandatory to calculate the error signal $\mathbf{e}^e$ (based on the quantized values $\mathbf{y}^e$ and b as in (3.117) with full precision arithmetic

$$\mathbf{e}^e(r) = b(r) \left[\mathbf{y}^{e*}(r) - \frac{\mathbf{y}^{e*}(r)^T \mathbf{c}^e}{L} \; \mathbf{c}^e \right]. \quad (3.117)$$

This means that, starting from quantized values $\mathbf{y}^e$ and b, the processing relevant to $\mathbf{e}^e$ (and so $\mathbf{x}^e$) has to be performed with an internal word length dictated by the whole signal dynamics, so that no further truncation is introduced. This reveals very demanding in terms of hardware complexity[2], but, as we show here for the simplified case of 1 bit quantization[3] of $\mathbf{x}^e(r)$,

---

[2]   The RAM memory needed to store the $3L$ components of $\mathbf{x}^e$ accounts for nearly half the overall EC-BAID silicon area.

[3]   The derivation is performed for simplicity in the case of *1* bit truncation, but can be easily generalized to *n* bit with $n > 1$.

has intolerable effects. Starting from (3.117) with $L = 128 = 2^7$, the internal 'full precision' (*f.p.*) representation of the error signal can be re-arranged as follows

$$\mathbf{e}^e(r)_{f.p.} = b(r)\left[\mathbf{y}^{e*}(r)2^7 - \left(\mathbf{y}^{e*}(r)^T \mathbf{c}^e \right)\mathbf{c}^e \right]$$ (3.118)

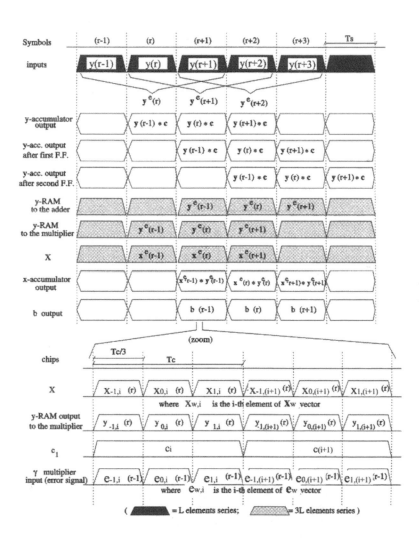

Figure 3-48. Timing diagram of the main signals for the 'Select and Add' architecture.

such that the 1 bit 'truncated' (*tr.*) version of the error signal is

$$\mathbf{e}^e(r)_{f.p.} = \left\lfloor b(r)\left[\mathbf{y}^{e*}(r)2^7 - \left(\mathbf{y}^{e*}(r)^T \mathbf{c}^e\right)\mathbf{c}^e\right]\right\rfloor_{tr}$$

$$= \mathbf{e}^e(r)_{f.p.} + \Delta\mathbf{e}^e(r)$$

(3.119)

where $\lfloor \ \rfloor_{tr}$ means 'zero forcing' of the LSB of $\mathbf{e}^e(r)_{f.p.}$.

Table 3-3. EC-BAID-I and -II equations for the 'Select and Add' architecture.

EC-BAID-I and –II detector output	$b(r) = \dfrac{1}{L}\ \mathbf{h}^e(r)^T\ \mathbf{y}^e(r)$ with $\mathbf{h}^e(r) = \mathbf{x}^e(r) + \mathbf{c}^e$
EC-BAID-I update of $\mathbf{x}_1^e$	$\mathbf{x}_w(r+1) = \mathbf{x}_w(r) - \gamma\mathbf{e}_w(r-1)$ $\mathbf{e}_w(r) = b(r)\left[\mathbf{y}_w^*(r) - \dfrac{\mathbf{y}_w^*(r)^T\mathbf{c}}{L}\mathbf{c}\right]\quad w = -1,0,1$
EC-BAID-II update of $\mathbf{x}_1^e$	$\mathbf{x}^e(r+1) = \mathbf{x}^e(r) - \gamma\mathbf{e}^e(r-1)$ $\mathbf{e}^e(r) = b(r)\left[\mathbf{y}^{e*}(r) - \dfrac{\mathbf{y}^{e*}(r)^T\mathbf{c}^e}{L}\mathbf{c}^e\right]$

$$\mathbf{c}^e = \begin{bmatrix} \mathbf{0} \\ \mathbf{c} \\ \mathbf{0} \end{bmatrix}, \quad \mathbf{x}^e(r) = \begin{bmatrix} \mathbf{x}_{-1}(r) \\ \mathbf{x}_0(r) \\ \mathbf{x}_1(r) \end{bmatrix}, \quad \mathbf{e}^e(r) = \begin{bmatrix} \mathbf{e}_{-1}(r) \\ \mathbf{e}_0(r) \\ \mathbf{e}_1(r) \end{bmatrix}$$

In (3.119), $\mathbf{e}^e(r)_{f.p.}$ is orthogonal to $\mathbf{c}^e$ by construction (i.e., $\mathbf{e}^e(r)_{f.p.}^T \mathbf{c}^e = 0$), whilst the same consideration does not apply to the quantization term $\Delta\mathbf{e}^e(r)$. In particular, taking into account that the LSB of the term ($\mathbf{y}^{e*}(r) \cdot 2^7$) in (3.119) is necessarily zero, $\Delta\mathbf{e}^e(r)$ can be expressed as follows

$$\Delta\mathbf{e}^e(r) = \mathbf{e}^e(r)_{tr.} - \mathbf{e}^e(r)_{f.p.}$$

$$= \left\lfloor -b(r)\left(\mathbf{y}^{e*}(r)^T\mathbf{c}^e\right)\mathbf{c}^e\right\rfloor_{tr} + b(r)\left(\mathbf{y}^{e*}(r)^T\mathbf{c}^e\right)\mathbf{c}^e.$$

(3.120)

To simplify the expression above, we let

$$\alpha(r) = b(r)\left(\mathbf{y}^{e*}(r)^T\mathbf{c}^e\right)$$

(3.121)

such that

$$\Delta\mathbf{e}^{e}(r)=\left\lfloor-\alpha(r)\mathbf{c}^{e}\right\rfloor_{tr}+\alpha(r)\mathbf{c}^{e} \tag{3.122}$$

Consider now that $\mathbf{c}^{e}$ is an extended vector whose central section with $w=0$ has components with values $+1$ or -1. The product with $\alpha(r)$ then gives two different possible results: if $\alpha(r)$ is even, 1 bit truncation does not introduce any error and we have

$$\left\lfloor-\alpha(r)\mathbf{c}^{e}\right\rfloor_{tr.}=-\alpha(r)\mathbf{c}^{e}\Rightarrow\Delta\mathbf{e}^{e}(r)=0; \tag{3.123}$$

if $\alpha(r)$ is odd, 1 bit truncation is equivalent, considering a 2's complement representation, to subtracting 1 to each non-zero vector element, thus

$$\left\lfloor-\alpha(r)\mathbf{c}^{e}\right\rfloor_{tr.}=-\alpha(r)\mathbf{c}^{e}-\left[00\cdots0\mid111\cdots1\mid00\cdots0\right]^{T}$$
$$\Rightarrow\Delta\mathbf{e}^{e}(r)=-\left[00\cdots0\mid111\cdots1\mid00\cdots0\right]^{T}. \tag{3.124}$$

The vector $\Delta\mathbf{e}^{e}(r)$ is made of a component which is orthogonal to $\mathbf{c}^{e}$ and of a component $\mathrm{d}\mathbf{e}^{e}$ which is not. The first will have no effect on the overall performance, whilst the second, being characterized by elements all of the same sign, will build up an accumulation error. This will impair algorithm convergence. In particular, $\mathrm{d}\mathbf{e}^{e}$ is given by

$$\mathrm{d}\mathbf{e}^{e}(r)=\left(\frac{\Delta\mathbf{e}^{e}(r)^{T}\mathbf{c}^{e}}{L}\right)\mathbf{c}^{e}=\left(\frac{\left[111\cdots1\right]^{T}\mathbf{c}}{L}\right)\mathbf{c}^{e}\neq0. \tag{3.125}$$

The term $([111\cdots1]^{T}\mathbf{c})$ in (3.125) is simply the sum of the code chips $c_{1}(\ell)$. So $\mathrm{d}\mathbf{e}^{e}$ is zero only if the sequence $\mathbf{c}$ is balanced, that is, it contains an equal number of $+1$ and -1. As previously stated, the MUSIC receiver supports the use of extended PN sequences overlaid to WH signatures. This superposition generates unbalanced codes for almost all of the possible combinations. Thus it is very likely, in the case of 1 bit truncation, to have $\mathrm{d}\mathbf{e}^{e}(r)\neq0$.

Figure 3-49 shows the estimated BER performance of the EC-BAID obtained with $L=128$, $K=64$, $E_{b}/N_{0}=5\,\mathrm{dB}$, $C/I=-6\,\mathrm{dB}$ and on a simulation run of 20 Msymbols. The lower (almost horizontal) curve was obtained with no truncation in the evaluation of $\mathbf{e}^{e}$, whilst the upper one was obtained with just 1 bit error in the internal word length dimensioning. In the latter case, the term $\left[111\cdots1\right]^{T}\mathbf{c}$ of (3.125) is equal to 16, and so every time

$\alpha(r)$ is odd, the component of $\mathbf{x}^e(r)$ which is not orthogonal to $\mathbf{c}^e$ is incremented by a 'drift' term $(16/128)\mathbf{c}^e$.

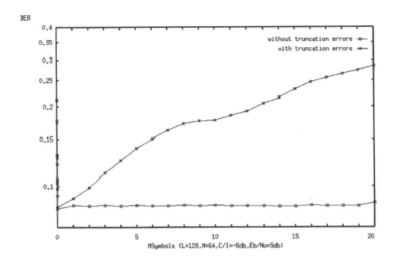

Figure 3-49. BER system performance with and without error signal truncation effects.

As is seen in Figure 3-47, an internal AGC on the EC-BAID feedback loop is needed in order to keep the output amplitude $|b|$ constant at different SNIR operating conditions. The amplitude of the received signal varies according to the amplitude of the different interfering user signals ($\sqrt{P_k}$), and according to the effects of the channel noise $n(mT_c)$. This is more clearly apparent if we assume synchronous multiplexing

$$y(m) = \sqrt{P_1}\,\tilde{a}_1(m/\!/L)c_1\left(|m|_L\right)$$
$$+ \sum_{k=2}^{K}\sqrt{P_k}\,\tilde{a}_k(m/\!/L)c_k\left(|m|_L\right) + n(m), \qquad (3.126)$$

where $n(m)$ is a Gaussian noise sample. In order to better exploit the input dynamic range of the ADC, the received signal is adjusted to a proper level by an IF Variable Gain Amplifier (VGA), which keeps the total signal power P at a constant level $P = \sigma_{ADC}^2 = A^2$. Considering a unit power useful channel (i.e., $P_1 = 1$ in (3.126)), the baseband signal is

$$y_{VGA}(m) = D\,y(m) = D\,\tilde{a}_1\left(\{m\}_L\right)c_1\left(|m|_L\right)$$
$$+ \sum_{k=2}^{K} D\,\sqrt{P_k}\,\tilde{a}_k\left(\{m\}_L\right)c_k\left(|m|_L\right) + D\,n(m), \qquad (3.127)$$

where D is the variable amplification of the VGA. It can be seen from Table 3-3 that the updating equation for $\mathbf{x}^e$ depends on y and b through the following term

$$\Delta \mathbf{x}^e = \gamma \mathbf{e}^e(r) = \gamma b(r) \left[\mathbf{y}^*(r) - \frac{\mathbf{y}^{e*}(r)^T \mathbf{c}^e}{L} \mathbf{c}^e \right]. \tag{3.128}$$

Considering the IF VGA, this term becomes

$$\Delta \mathbf{x}^e_{\text{VGA}} = \gamma b_{\text{VGA}}(r) \left[\mathbf{y}^{e*}_{\text{VGA}}(r) - \frac{\mathbf{y}^{e*}_{\text{VGA}}(r)^T \mathbf{c}^e}{L} \mathbf{c}^e \right] = D^2 \Delta \mathbf{x}^e \tag{3.129}$$

because the factor D affects both $b_{\text{VGA}} = D\,b$ and $\mathbf{y}^e_{\text{VGA}} = D\,\mathbf{y}^e$. So, with the same value of γ, the amplitude of $\Delta \mathbf{x}^e$, and thus the algorithm convergence speed, is affected by D, thus is in turn determined by the actual SNIR conditions. Considering the MUSIC signal requirements, we derive the following range for P:

i) maximum SNIR condition:
$E_c/(N_0 + I_0) \to \infty$, $K = 1$ (i.e. $y = \sqrt{P_1}\, \tilde{a}_1 c_1$)
$\Rightarrow |y|^2 \approx P_1 = 1 \Rightarrow D \approx A = D_{\text{max}}$;

ii) minimum SNIR condition:
$E_c/(N_0 + I_0) = -25$ dB
$\Rightarrow |y|^2 \approx 315 \cdot P_1 \Rightarrow D \approx A/18 = D_{\text{min}} \Rightarrow D^2_{\text{max}} / D^2_{\text{min}} = 315$.

This means that for the same value of γ the acquisition time can vary considerably, and this is also why a suited local digital amplitude control has to be implemented within the S&A architecture. In particular, Figure 3-50 shows the block diagram of the digital AGC in Figure 3-47. The circuit (which is rather simplified, since its main purpose is just to keep the acquisition time roughly constant) computes on a proper time window the average value $|b_{\text{VGA}}|$ of the quantity

$$|b_{\text{VGA}}| = \frac{|b_{\text{VGA},P}| + |b_{\text{VGA},Q}|}{2} \tag{3.130}$$

and scales accordingly the b_{VGA} signal through the selector. The effect of the AGC on the EC-BAID convergence speed is clearly shown in Figures 3-51 and 3-52, where we show the evolution of the BER with time for two different SNIR scenarios, with and without AGC. Equalization of the convergence speed is apparent.

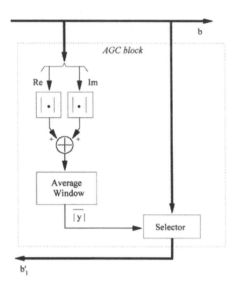

Figure 3-50. Digital AGC circuit.

In practice the averaging of the instantaneous amplitude (3.130) is performed by the simple recursive circuit in Figure 3-53, where the amplitude $|b(r)|$ of the EC-BAID output is actually the "simplified" amplitude (3.130).

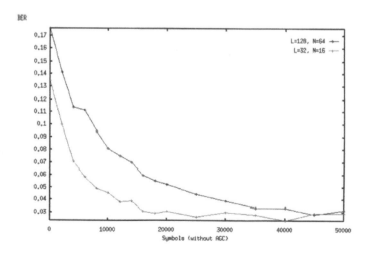

Figure 3-51. BER transient for two different SNIR scenarios without digital AGC.

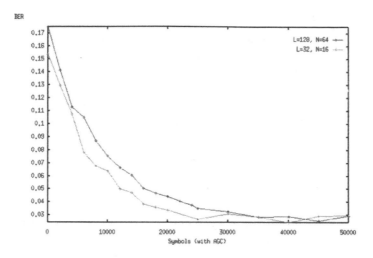

Figure 3-52. BER transient for two different SNIR scenarios with digital AGC.

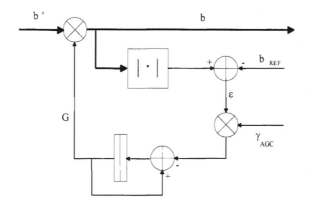

Figure 3-53. EC-BAID Internal (digital) AGC.

The operations performed in Figure 3-53 are summarized as follows

$$b_{\text{OUT}}(r) = G(r)b_{\text{IN}}(r), \tag{3.131}$$

$$\varepsilon(r) = |b(r)| - b_{\text{REF}}, \tag{3.132}$$

$$G(r+1) = G(r) - \gamma_{\text{AGC}}\varepsilon(r). \tag{3.133}$$

The value of the loop constant γ_{AGC} was set in the range $2^{-4} \div 2^{-5}$. After some preliminary trials this value was found to be large enough so that the evolution of the AGC is always faster than that intrinsic to the EC-BAID.

4.   RECEIVER ARCHITECTURE AND SIMULATION RESULTS

Taking into consideration the outcomes of the design study carried out in the previous Sections, the final architecture of the MUSIC detector is the one depicted in Figure 3-54.

Such architecture was the subject of extensive simulations to test the joint behavior of all of the subsystems described up to now. The Numerical results we are going to present in the sequel concern both Floating Point (FP) and Bit True (BT) simulations of the overall receiver.

4.1   Floating Point Simulations and Architectural Settings

The baseline configuration for the MUSIC receiver was optimized in the following reference scenario:

i)   asynchronous equi-power interferers, with time delays evenly spaced within one symbol interval and phase shifts evenly spaced within 2π;

ii)   Extended Gold overlay sequence;

iii) pilot signal code 0 out of the Walsh–Hadamard set of orthogonal sequences;

iv)   useful traffic signal code 5 out of the Walsh–Hadamard set of orthogonal sequences;

v)   useful traffic time and phase synchronous, and orthogonal with respect to the pilot signal;

vi)   pilot signal with no data modulation;

vii) useful traffic signal to single interferer power ratio $C/I = -6\,\text{dB}$

viii) pilot to useful traffic signal power ratio $P/C = 6\,\text{dB}$

ix)   EC-BAID observation window $W_{LEN} = 2$ (shortened with respect of the standard window $W_{LEN} = 3$, see below);

x)   internal AGC EC-BAID active with adaptation coefficient $\gamma_{AGC} = 2^{-4}$.

In our simulations the EC-BAID step size γ_{BAID} was allowed to take values between 2^{-13} and 2^{-15}, whilst, according to the receiver specifications, the spreading factor was 32, 64, or 128.

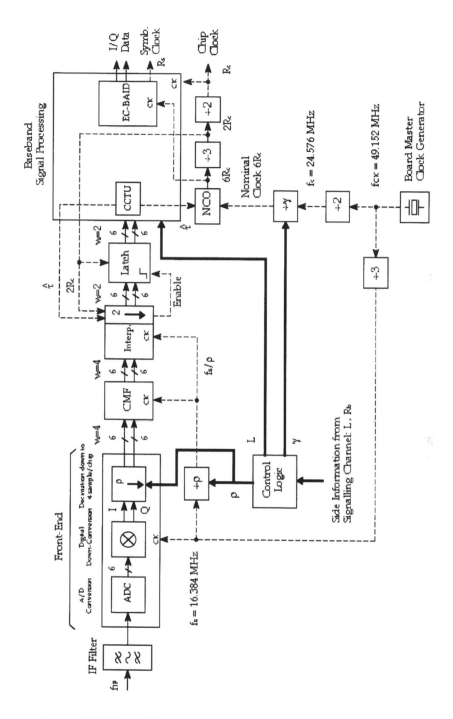

Figure 3-54. High level block diagram of the EC-BAID-based CDMA receiver, with some implementation details.

Regardless of its value, a half-loading condition, with $K = L/2$ active users, was typically investigated as the worst case. The BER performance of the EC-BAID was derived for different widths W_{LEN} of the observation window (normalized with respect to the symbol interval) ranging from 1 (as in the case of the conventional MMOE receiver) to 3 (which is the extended window length already investigated during previous studies on the BAID detector [De98a], [Rom97] and presented in Section 3.1).

From numerical results it turned out that the optimum choice for the detector observation window lies somewhere in the vicinity of the value $W_{LEN} = 2$. We therefore set the window length equal to *two* symbol intervals centered on the symbol period under estimation. Actually, the window length $W_{LEN} = 3$ of the original EC-BAID [Rom97] was found to sometimes introduce excess noise.

All simulations were allowed to run for a minimum number of 50 ksymbols, after completion of the initial training of the EC-BAID. Depending on the adaptation step γ_{BAID}, 20 or 50 ksymbols were allowed for algorithm convergence (training), and were not taken into account for BER computations. The ratio between training and valid data transmission length is reported in all charts.

Once the architecture of the EC-BAID was settled, we evaluated first the impact of the filtering/decimation scheme implemented in the MRFEU.

As a sample of this analysis Figure 3-55 shows our BER results with the system configuration summarized in the chart inset. The decimation factor of the front end filters is $\rho = 8$, and ideal chip timing and carrier frequency and phase recovery are considered. The performance degradation entailed by the introduction of the MRFE is immaterial in spite of the presence of several (e.g., 32) interferers. This behavior was confirmed for every downsampling ratio.

To properly design the code timing tracking loop in the CCTU, we evaluated the sensitivity of the EC-BAID to a chip timing error, in order to start designing the recovery loop with good initial guesses about the required loop bandwidth.

To this end, the BER performance of the BAID detector was derived in the presence of a chip sampling jitter. The chip clock jitter was modeled as a zero mean correlated Gaussian random process, with normalized variance σ^2 and with bandwidth $B = 10^{-3}/T_c$.

Numerical results were produced for different values of the normalized jitter variance σ^2, with $C/I = -6$ dB and $L = 64$. As usual, the interferers' time delays and carrier phase shifts were set uniformly spaced in the intervals $[0, T_s]$ and $[0, 2\pi]$, respectively.

The outcome of this analysis was that the detector is robust against chip clock jitter with normalized variances up to 5×10^{-3}.

The maximum tolerable RMS value of the CCTU estimate, normalized with respect to the chip interval, is then $\sigma = 0.071$.

Therefore, we set as our design goal for the CCTU an RMS chip timing error $\sigma < 0.05$ (less than 5% of a chip).

This was found to be the major drawback of the EC-BAID, especially when compared to the conventional (data aided) MMSE detector which is insensitive to (a small) timing jitter.

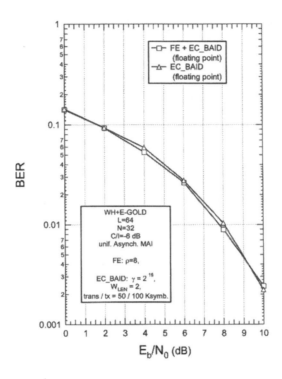

Figure 3-55. Impact of MRFE on BER performance.

Figure 3-56 shows the BER performance of the EC-BAID equipped with the MRFEU and the CCTU with step size $\gamma_{CCTU} = 2^{-7}$. The SACU adaptation step size was set to $\gamma_{AGC} = 2^{-5}$ in order to have fast AGC acquisition. Ideal carrier frequency/phase synchronization was still considered for this run. As is seen, the impact of the CCTU is non-negligible only at the highest values of SNR. The relevant degradation is always smaller than 0.5 dB, consistent with the results about the sensitivity to chip timing jitter addressed above. The value $\gamma_{CCTU} = 2^{-7}$ roughly corresponds to a normalized noise loop bandwidth equal to $B_n \approx 2^{-7} / T_s$. In a first stage of design we also tried to optimize the architecture of the CCTU in order to speed up the convergence of the algorithm.

We considered then a solution using two different step sizes in the two phases of loop *Acquisition* ('ACQ') and *Steady State* ('SS') tracking, denoted as γ_{ACQ} and γ_{SS}, respectively. After preliminary investigation, it was found that a sufficiently large value of the step size to have reasonably fast acquisition is $\gamma_{ACQ} = 2^{-5}$. However, the CCTU acquisition transient was always negligible with respect to the one of the EC-BAID or the AFCU, thus we eventually kept the same step size for both acquisition and steady state tracking, i.e., $\gamma_{ACQ} = \gamma_{SS} = \gamma_{CCTU} = 2^{-7}$. It is known that a worst case for the CCTU to operate in, is the condition of *synchronous* (quasi-) orthogonal interferers in the forward link, owing to the bad off sync spurious cross-correlation peaks of WH functions. We tested the performance of our CCTU in such condition and we found about 0.3 dB degradation introduced by MRFEU, and further 0.2 dB only caused by CCTU.

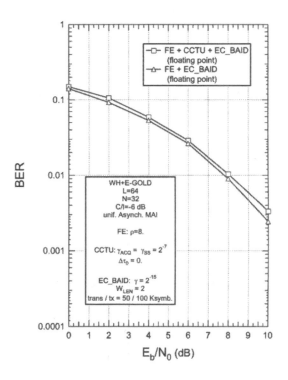

Figure 3-56. Impact of CCTU on BER performance.

Finally, Figure 3-57 summarizes the outcome of a thorough overall FP receiver simulation also including AFCU and CPRU. In particular, the AFCU step size was set to $\gamma_{AFCU} = 2^{-19}$ and the initial frequency error to zero, whilst for the CPRU we set $\gamma_{CPRU} = \rho_{CPRU} = 2^{-9}$. As is apparent from Figure 3-57, the carrier frequency/phase loops barely affect the system

performance in terms of BER. The reference curve labeled 'EC-BAID' is
obtained with all sync parameters perfectly recovered.

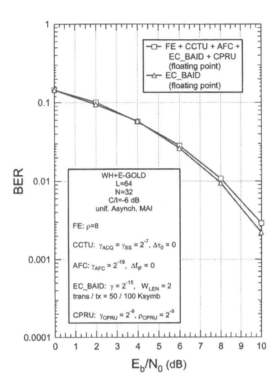

Figure 3-57. FP BER performance of MUSIC receiver.

In the overall receiver simulation, the acquisition phase was prudentially
set at 50,000 symbols, substantially owing to the AFCU acquisition time
(Figure 3-33), although in a real environment and when the SNR is good, the
receiver starts bearing a BER lower than 1% much earlier. It was also found
that the degradation caused by sync subsystems and by the MRFEU is
noticeable only with $L = 32$, wherein probably the small spreading factor is
not enough to sufficiently 'decorrelate' the additional noise terms owing to
the diverse degradation factors. The SNR degradation is anyway smaller
than 1 dB in any tested configuration, and is particularly modest when the
BER is around 1% (target value for unprotected voice transmission). We
remark that these findings are relevant to FP simulations, and should be
verified in Section 4.2 against BT results obtained after signal quantization
and finite arithmetic effects have been introduced. Offline with respect to the
steady state simulations reported above, we also performed analysis and

design of the CTAU. The outcome of such investigation was that, in order to
meet the performance requirements, the CTAU settings must be as follows:

i) $W = 512$ or 1024 ;
ii) $\lambda = 1.7$ or 1.8 ;
iii) $M = 1$;
iv)   non-orthogonal pilot.

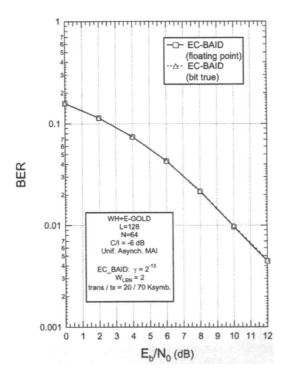

Figure 3-58. EC-BAID BT performance

Such a selection of the CTAU parameters was demonstrated to yield:

i)   probability of False (signal) Detection (P_{FD}) lower than 0.001;
ii)   probability of Missed (signal) Detection (P_{MD}) lower than 0.001;
iii) probability of Wrong (code phase) Acquisition (P_{WA}) lower than 0.001.

4.2   Quantization and Bit True Performance

The final step in the verification of our design was evaluation of the
effects of internal quantization on the BER performance. We started

considering the performance degradation of the EC-BAID with the following bit true configuration of the I/O and main internal signals:

i) input I/Q signals: 7 bits;
ii) internal vector $\mathbf{x}^e$: 10 bits/component;
iii) internal register for update of vector $\mathbf{x}^e$: 23 bits/component;
iv) AGC gain: 8 bits.

Figure 3-58 shows the effects of quantization on the EC-BAID performance, considering ideal front end filtering and chip timing tracking as well as ideal carrier frequency/phase recovery.

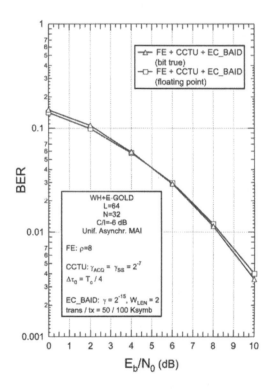

Figure 3-59. BER performance.

We can appreciate that the SNR degradation of the BT EC-BAID with respect to its ideal FP counterpart is always smaller than 0.5 dB, except for the worst case of $L = 32$ and $N = 16$ (not reported here) at the highest SNRs, where the BER is anyway sufficient. In this bit true run the received baseband signal is 'artificially' digitized at baseband and at chip rate in front of the (baseband) EC-BAID with an optimum setting. The agreement of these BT results with those actually got from ADC carried out at IF was later checked, and found satisfactory. Figure 3-59 shows the BER curves of the EC-BAID detector now equipped with BT MRFE and LIU/CCTU, but still considering ideal carrier frequency/phase recovery. The front end decimation factor is $\rho = 8$, whilst the CCTU step size[4] and initial timing error are $\gamma_{CCTU} = 2^{-7}$, $\tau_0 = T_c/4$, respectively. The ADC operates on 8 bits, and this input/output word length is kept in the subsequent operations of baseband conversion and CIC filtering[5]. In contrast, the CMF output (LIU input) is efficiently represented by 7 bits. The bit true configuration of the LIU and the CCTU is summarized as follows:

i)   7 bit LIU input/output with fractional delay μ represented on 5 bit;
ii)   9 bit representation of the (normalized) timing error ε;
iii) 7 bit input/output SACU signals with 7 bit gain factor ρ_{AGC}.

   The design verification is concluded by evaluation of the impact of the carrier recovery subsystems. Figure 3-60 shows the comparison between the BER of an overall bit true system simulation with ideal (triangles) and actual (dots) carrier recovery (AFCU+CPRU). The specific loss owed to carrier synchronization is negligible as long as the BER is lower than 0.1. Instead, at low SNRs the loop noise may result is a significant impairment of performance. Once again independently from steady state optimization, bit true performance of the CTAU with asynchronous interference was also tested. Here the crucial factor to reduce implementation complexity is signal quantization in the front end correlator. The impact of front end quantization (1, 2, or 3 bits) was found non-negligible with respect to ideal FP simulations, but it was also seen that the performance degradation owed to raw quantization can be anyway compensated through appropriate lengthening of the smoother window length *W*. In the limit even 1 bit quantization (i.e., consideration of the MSB of the MRFE output) may turn out to meet the acquisition specifications of the MUSIC receiver. This is why our final choice for the CTAU architecture was just 1 bit quantization (and, as outlined in the previous section, 1-symbol coherent correlation only). Figure 3-61 compares curves for P_{WA} with FP and BT MRFEU in the two extreme cases of oversampling ratio $\rho = 2$ and $\rho = 32$, and assuming ideal chip epoch recovery. The front end quantization in the correlator is always 1 bit.

---

[4]   The effect of the digital SACU in Figure 3-2 is to keep the maximum amplitude of the CCTU S curve close to unity in every operating condition. The slope of the S curve is therefore roughly equal to 4, and the normalized loop noise bandwidth is $B_n T_s = \gamma_{CCTU}$.

[5]   This is an input/output figure. Internal processing should occasionally be done with larger word lengths.

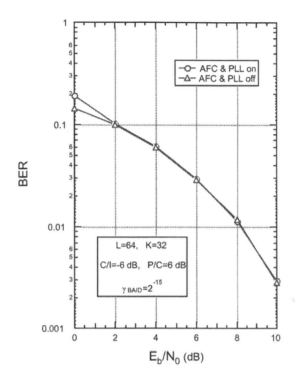

Figure 3-60. Influence of AFCU and CPRU on EC-BAID BT performance.

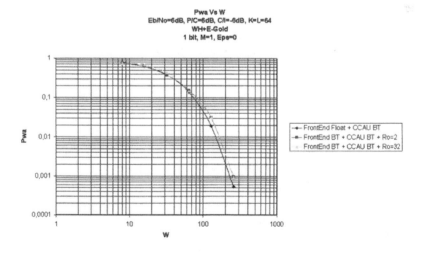

Figure 3-61. Comparison between FP front end and BT front end ($L = 64$).

Chapter 4

FROM SYSTEM DESIGN TO HARDWARE PROTOTYPING

After the previous chapter the reader should have quite clear in their mind the main architectural solutions of the different signal detection issues which were highlighted. The question now is how to translate it into good hardware design. Introduced by a brief discussion about the main issues in design and implementation of wireless telecommunication terminals (design flows, design metrics, design space exploration, finite arithmetic effects, rapid prototyping, etc.), this Chapter presents in detail the FPGA hardware implementation of the CDMA receiver described in Chapter 3.

1. VLSI DESIGN AND IMPLEMENTATION OF WIRELESS COMMUNICATION TERMINALS: AN OVERVIEW

As discussed in Chapter 1, the only viable solution for handling both the exponentially increasing algorithmic complexity of the physical layer and the battery power constraint in wireless terminals is to rely on a heterogeneous architecture which optimally explores the 'flexibility–power–performance–cost' design space. In this respect Figure 1-14 in Chapter 1 shows a typical heterogeneous *System on a Chip* (SoC) architecture employing several programmable processors (either standard and application specific), on chip memories, bus based architectures, dedicated hardware coprocessors, peripherals and I/O channels. The current trend in the design of digital terminals for wireless communications consists in moving from the integration of different physical components in a system printed circuit board to the integration of different virtual components[1] in a SoC.

As far as computational processing is concerned, we can identify three typical digital 'building blocks' which are characterized by different 'energy–flexibility–performance' features: microprocessors, general purpose *digital signal processors* (DSPs) and *application specific integrated circuits* (ASICs).

A fully programmable microprocessor is better suited to perform the non-repetitive, control oriented, input/output operations, as well as all the house-keeping tasks (such as protocol stacks, system software and interface software). Embedded micro cores are provided by ARM [arm], MIPS [mips], Tensilica [tensi], IBM [ibm], ARC [arc] and Hitachi [hitac], just to name a few.

Programmable DSPs are specialized VLSI devices designed for implementation of extensive arithmetic computation and digital signal processing functions through downloadable, or resident, software/firmware. Their hardware and instruction sets usually support real time application constraints. Classical examples of signal processing functions are *finite impulse response* (FIR) filters, the *Fast Fourier Transform* (FFT), or, for wireless applications, the *Viterbi Algorithm* (VA). We notice that conventional (general purpose) microprocessors, although showing significantly higher power consumptions, do not generally include such specialized architectures. DSPs are typically used for speech coding, modulation, channel coding, detection, equalization, or frequency, symbol timing and phase synchronization, as well as amplitude control. Amidst the many suppliers of embedded DSP cores, we mention here STMicroelectronics [stm], Motorola [motor], Lucent [lucen] and Texas Instrument [ti].

A DSP is also to be preferred in those applications where flexibility and addition of new features with minimum re-design and re-engineering are at a premium. Over the last few years, the pressure towards low power consumption has spurred the development of new DSPs featuring hardware accelerators for Viterbi/Turbo decoding, vectorized processing and specialized domain functions. The combination of programmable processor cores with custom accelerators within a single chip yields significant benefits such as performance boost (owing to time critical computations implemented in accelerators), reduced power consumption, faster internal communication between hardware and software, field programmability owed to the programmable cores and, last but the least, lower total system cost owed to the single-DSP chip solution.

---

[1] A 'virtual component' is what we may call an *intellectual property* (IP) silicon block. The *Virtual Socket Interface* (VSI) Alliance was formed in 1996 to foster the development and recognition of standards for designing re-usable IP blocks [vsi].

ASICs are typically used for high throughput tasks in the area of digital filtering, synchronization, equalization, channel decoding and multiuser detection. In modern 3G handsets the ASIC solution is also required for some multimedia accelerators such as the *Discrete Cosine Transform* (DCT) and *Video Motion Estimation* (VME) for image/video coding and decoding. From an historical perspective, ASICs were mainly used for their area–power efficiency, and are still used in those applications where the required computational power could not be supported by current DSPs.

Thanks to the recent advances in VLSI technology the three 'building blocks' we have just mentioned can be efficiently integrated into a single SoC. The key point remains how to map algorithms onto the various building blocks (software and hardware) of a heterogeneous, configurable SoC architecture. The decision whether to implement a functionality into a hardware or software subsystem depends on many (and often conflicting) issues such as algorithm complexity, power consumption, flexibility/programmability, cost, and time to market. For instance, a software implementation is more flexible than a hardware implementation, since changes in the specifications are possible in any design phase. As already mentioned in Chapter 1, a major drawback is represented by the higher power consumption of SW implementations as compared to an ASIC solution, and this reveals a crucial issue in battery operated terminals. For high production volumes ASICs are more cost effective, though more critical in terms of design risk and time to market. Concerning the latter two points, *computer aided design* (CAD) and system-level tools enabling efficient algorithm and architecture exploration are fundamental to turning system concepts into silicon rapidly, thus increasing the productivity of engineering design teams.

1.1   Simplified SoC Design Flow

A typical design flow for the implementation of an algorithm functionality into a SoC, including both hardware and software components, is shown in Figure 4-1. The flow encompasses the following main steps:

1. creation a system model according to the system specification;
2. refinement of the model of the SoC device;
3. hardware–software partitioning;
4. hardware–software co-simulation;
5. hardware–software integration and verification;
6. SoC tape out.

The first step consists in modeling the wireless system (communication transmitter and/or receiver, etc.) of which the SoC device is part of. Typically, a floating point description in a high level language such as MAT-LAB, C or C++ is used during this phase. Recently there has been an important convergence of industry/research teams onto SystemC[2] as the leading approach to system level modeling and specification with C++.

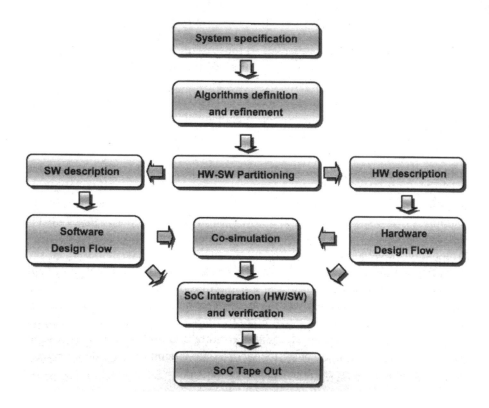

Figure 4-1. Simplified SoC Design Flow.

Today most *electronic design automation* (EDA) suppliers support SystemC. Within such a programming/design environment, high level *intellectual property* (IP) modules being commercially available helps to boost design efficiency and verifying compliance towards a given reference standard. Based on these IPs designers can develop floating point models of digital modems by defining suitable algorithms and verifying performance via system level simulations. The system model is firstly validated against well known results found in the literature as well as theoretical results (BER curves, performance bounds, etc.) in order to eliminate possible modeling or

---

[2]   The rationale behind the Open SystemC Initiative [syste] is to provide a modeling framework for systems where high level functional models can be refined down to implementation in a single language.

simulation errors. Simulations of the system model are then carried out in order to obtain the performance of a 'perfect' implementation, and consequently to check compliance with the reference standard specification (i.e., 2G, 3G, etc.). The outcomes of this second phase are considered as the benchmark for all successive design steps which will lead to the development of the final SoC algorithms. Currently many design tools for system simulation are available on the market, such as CoCentric System Studio™ and COSSAP™ by Synopsys [synop], SPW™ by Cadence [caden], MAT-LAB™ by MathWorks [mathw], etc.. The legacy problem and high costs often slow down the introduction of new design methodologies and tools. Anyway, different survey studies showed that the most successful companies in the consumer, computer and communication market are those with the highest investments in CAD tools and workstations.

Following the phase of system simulation, joint algorithm/architecture definition and refinement takes place. This step, which sets the basis for hardware/software partitioning, typically includes the identification of the parameters which have to be run time configurable and those that remain preconfigured, the identification (by estimation and/or profiling) of the required computational power (typically expressed in number of *operations per second* — OPs), and the estimation of the memory and communication requirements. The partitioning strategy not only has a major impact on die size and power consumption, but also determines the value of the selected approach for re-use in possible follow up developments. In general, resorting to dedicated building blocks is helpful for well known algorithms that call for high processing power and permanent utilization (FFT processors, Turbo decoding, etc.). The flexibility of a DSP (or micro) core is required for those parts of a system where complexity of the control flow is high, or where subsequent tuning or changes of the algorithms can achieve later market advantages or an extension of the SoC application field.

After partitioning is carried out the (joint) development of hardware and software requires very close interaction. Interoperability and interfacing of hardware and software modules must be checked at any stage of modeling. This requires co-simulation of the DSP (or micro) processor *instruction set* (IS) with the dedicated hardware. Once a dream, co-simulation is nowadays a reality for many processors within different CAD products available on the market, such as Synopsys [synop], Cadence [caden], Coware [cowar] and Mentor Graphics [mento]. In particular, finite word length effects have to be taken into account in both hardware and software modules by means of bit true simulation. This requires the conversion of the original model from floating to fixed point. Such a process reveals a difficult, error prone and time consuming task, calling for substantial amounts of previous experience, even if support from CAD tools is available (such as, for instance, the Co-

Centric System Studio™ Fixed Point Designer by Synopsys). Thus the final system performance can be assessed, the actual *implementation loss*[3] can be evaluated. Even though the algorithms are modified from the original floating point model, the interfaces of the SoC model are kept. The bit true model can always be simulated or compared against the floating point one, or it can be simulated in the context of the entire system providing a clear picture of the tolerable precision loss in the fixed point design.

Overall system simulation is particularly relevant when different building blocks have to evaluated jointly to assess overall performance, and no separate requirements for the building blocks are provided. In cellular mobile communications systems absolute performance limits are given in terms of conformance test specifications, which indicate certain tests and their corresponding results boundaries. However, standards generally specify only *overall* performance figures. Let us consider, for instance, a specification for the *block error rate* (BLER) at the output of the channel decoder, whose performance depends on the entire physical layer (analog front end, digital front end, modem, channel decoder, etc.). The standard does not provide modem or codec specifications, but only overall performance tests. Thus no absolute performance references or limits exist for the major sub-blocks that can be used in the design process. This situation can be successfully tackled by starting with floating point models for the sub-blocks. These models can be simulated together to ascertain whether they work as required, and a tolerable implementation loss with respect to the floating point model can then be specified as the design criterion for the fixed point model. The final model serves then as an executable bit true specification for all the subsequent steps in the design flow.

Software design flow for DSP processor typically assumes throughput and RAM/ROM memory requirements as key optimization criteria. Unfortunately, when implementing complex and/or irregular signal processing architectures, even the latest DSP compilers cannot ensure the same degree of optimization that can be attained by the expert designer's in depth knowledge of the architecture. As a result, significant portions of the DSP code

---

[3] Two main issues must to be considered when dealing with finite word lengths arithmetics: (i) each signal sample (which is characterized by infinite precision) has to be approximated by a binary word, and this process is known as *quantization*; (ii) it may happen that the result of a certain DSP operation should be represented by a word length that cannot be handled by the circuit downstream, so the word length must be reduced. This can be done either by *rounding*, by *truncation*, or by *clipping*. The finite word length representation of numbers in a wireless terminal has ideally the same effect as an additional white noise term and the resulting decrease in the signal to noise ratio is called the *implementation loss* [Opp75]. For hardware dedicated logic the chip area is, to a first approximation, proportional to the internal word length, so the bit true design is always the result of performance degradation and area complexity trade off.

need to be tuned by hand (to explicitly perform parallelization, loop unrolling, etc.) to satisfy the tight real time requirements of wireless communications. Of course, this approach entails many drawbacks concerning reliability and design time. In this respect, DSP simulation/emulation environment plays an important role for code verification and throughput performance assessment.

Once a bit true model is developed and verified, the main issue in the hardware design flow is to devise the optimum architecture for the given cost functions (speed, area, power, flexibility, precision, etc.) and given technology. This is usually achieved by means of multiple trade offs: parallelism vs. hardware multiplex, bit serial vs. bit parallel, synchronous vs. asynchronous, precision vs. area complexity etc.. First, the fixed point algorithms developed at the previous step are refined into a cycle true model, the latter being much more complex than the former, and thus requiring a greater verification effort. Refining the fixed point model into a cycle true model involves specifying the detailed HW architecture, including pipeline registers and signal buffers, as well as the detailed control flow architecture and hardware–software interfaces. This final model serves as a bit- and cycle true executable specification to develop the *hardware description language* (HDL) description of the architecture towards the final target implementation.

Many different HW implementation technologies such as FPGA (*field programmable gate array*), gate array, standard cell and full custom layout are currently available. From top to bottom, the integration capability, performance, non-recurrent engineering cost, development time, and manufacturing time increase, and cost per part decreases owing to the reduced silicon area. The selection of the technology is mainly based on production volume, required throughput, time to market, design expertise, testability, power consumption, area and cost trade off. The technology chosen for a certain product may change during its life cycle (e.g., prototype on several FPGAs, final product on one single ASIC). In addition to the typical standard cells, full custom designed modules are generally employed in standard cell ICs for regular elements such as memories, multipliers, etc. [Smi97].

For both cell based and array based technology an ASIC implementation can be efficiently achieved by means of logic synthesis given the manufacturer cell library. Starting from the HDL (typically IEEE Std. 1076 – VHDL and/or IEEE Std. 1364 Verilog HDL) system description at the *register transfer level* (RTL), the synthesis tool creates a netlist of simple gates from the given manufacturer library according to the specified cost functions (area, speed, power or a combination of these). This is a very mature field and it is very well supported by many EDA vendors, even if Synopsys

Design Compiler™, which has been in place for almost two decades, is currently the market leader.

In addition to CAD tools supporting RTL based synthesis, some new tools are also capable of supporting direct mapping to cell libraries of a behavioral description. Starting from a behavioral description of the function to be executed, their task is to generate a gate level netlist of the architecture and a set of performance, area, and/or power constraints. This allows the assessment of the architectural resources (such as execution units, memories, buses and controllers) that are needed to perform the task (*allocation*), binding the behavioral operations to hardware resources (*mapping*), and determining the execution order of the operations on the produced architecture (*scheduling*). Although these operations represent the core of behavioral synthesis, other steps, for instance such as *pipelining*, can have a dramatic impact on the quality of the final result. The market penetration of such automated tools is by now quite limited, even if the emergence of SystemC as a widely accepted input language might possibly change the trend [DeM94].

After gate level netlist generation, the next step taking place is physical design. First, the entire netlist is partitioned into interconnected larger units. The placement of these units on the chip is then carried out using a floor planning tool, whilst a decision about the exact position of all the cells is done with the aid of placement and routing tools. The main goal is to implement short connection lines, in particular for the so called *critical path*. Upon completion of placement, the exact parameters of the connection lines are known, and a timing simulation to evaluate the behavior of the entire circuit can be eventually carried out (*post layout simulation*). Whether not all requirements are met, iteration of the floor planning, placement and routing might be necessary. This iterative approach, however, has no guarantee of solving the placement/routing problem, so occasionally an additional round of synthesis must be carried out based on specific changes at the RTL level. Once the design is found to meet all requirements, a programming file for the FPGA technology, or the physical layout (the GDSII format binary file containing all the information for mask generation) for gate array and standard cell technologies will be generated for integration in the final SoC [Smi97]. Finally, SoC hardware/software integration and verification, hopefully using the same testbench defined in the previous design steps, takes place and then *tape out* comes (the overall SoC GDSII file is sent out to the silicon manufacturer).

Very often rapid prototyping is required for early system validation and software design before implementing the SoC in silicon. Additionally, the prototype can serve as a vehicle for testing complex functions that would otherwise require extensive chip level simulation. Prototypes offer a way of emulating ASICs in a realistic system environment. Indeed, wireless systems often have very stringent *Bit Error Rate* (BER) requirements. For example, the typical BER requirement for a 2G system is approximately 10^{-2} (voice communications), whereas it may be as low as 10^{-6} (multimedia) for a 3G system. In general, the lower the BER requirements, the longer must be the bitstream to be simulated to achieve statistically valid results[4]. As a rule of the thumb we can assume that, in the case of randomly distributed errors, a reliable estimate of the BER with the error counting technique can be obtained by observing about 100 error events. It follows that in order to reliably measure a BER of 10^{-2}, about 10^4 symbols must be simulated, while a BER of 10^{-6} requires about 10^8 symbols. This can be unfeasible especially for verification at the lowest level of abstraction. Many rapid prototyping environment are available on the market for system emulation (such as Cadence [Smi97], Aptix [aptix], FlexBench [Pav02], Nallatech [nalla] and Celoxica [celox]). Alternatively, a prototyping environment can be developed in house, exploiting FPGA technology, possibly with downgrading of speed performance with respect to an ASIC solution, but still validating the logic functioning and hardware/software interfaces. Basing the FPGA prototype development exclusively on ASIC design rules, makes FPGA to ASIC technology conversion unnecessary, and lets the design version verified in the prototype ready for ASIC SoC implementation.

The following Sections of this Chapter present the design of the all-digital MUSIC receiver for hardware emulation, based on a custom designed platform. Particularly, rapid prototyping on FPGA technology for the EC-BAID ASIC is presented. The relevant ASIC design flow for a 0.18 μm CMOS standard cell technology will be detailed in Chapter 5.

2. FPGA IMPLEMENTATION OF THE ALL DIGITAL MUSIC RECEIVER

Following the general design rules outlined in the previous Section, the final architecture of the MUSIC receiver as in Section 3.4 was simulated in a high level general purpose programming language. For legacy reasons the scientific computation language FORTRAN was used, but the same results would have been obtained with C or C++. Through this simulator, or through relevant subsections, the different receiver subsections were designed and optimized as detailed in Chapter 3.

---

[4] All of the considerations reported here about BER estimation by means of measurement on the hardware prototype refer to the simple error counting technique (also addressed to as Monte Carlo method) which evaluates the error probability as the ratio between the number of observed errors and the number of transmitted bit, within a given time interval.

After that, the bit true, fixed point architecture of the receiver was simulated by means of a parametric FORTRAN model derived from the above-mentioned floating point simulation. The bit true model allowed determination of the wordlength of all internal digital signal as a trade off between complexity and overall performance. Bit true and floating point performances were continually compared to satisfy the given constraint of a maximum degradation of 0.5 dB. Once this goal was achieved, the circuit was described at the *Register Transfer Level* (RTL) with the VHDL (*Very high speed integrated circuit Hardware Description Language*) hardware description language, and the resulting model was input to the subsequent logic synthesis stage. The receiver was also equipped with extra auxiliary modules for monitoring and control. This allowed final evaluation and verification of the HW by means of direct comparison with the expected simulated results. This debugging activity will be detailed later in Chapter 6.

FPGA implementation represents the final goal of the receiver front end and synchronization loops. In contrast, it is only an intermediate phase for the EC-BAID design — it is just the stage of fast prototyping before ASIC implementation. Rapid prototyping aims at validating the system architecture before submission of the physical layout to the foundry. Therefore, the EC-BAID was described in VHDL as an ASIC core, and such circuit was directly targeted to FPGA technology without any modifications. This entailed a certain downgrading of speed performance: the FPGA implementation of the EC-BAID circuit could properly work for a subset of the required chip rates only, specifically from 128 kchip/s to 512 kchip/s. No pipeline registers were added to speed up the FPGA clock frequency, since the goal of the prototyping was testing the ASIC RTL with no changes.

A summary of the digital design flow that led to the FPGA implementation of the MUSIC receiver is sketched in Figure 4-2. This is conceptually very close to what described in the previous Section, and almost identical to the one that will be detailed in Chapter 5 for the ASIC implementation, with the only exception of the target technology. As a general rule, it is good practice in creating the design for the ASIC, first to verify and test it, and only then to implement the changes necessary for translating the design to FPGA technology. Operating the other way round (from FPGA design to ASIC) is more risky. First, errors in the translation are not visible in the prototype, and thus are not revealed in prototype testing. Second, the test structures for ASIC (Scan Path, memory BIST, etc.) are not implemented in the native design for FPGA. When the design is ported to the ASIC the test structures need to be added and re-verified with another iteration on the FPGA.

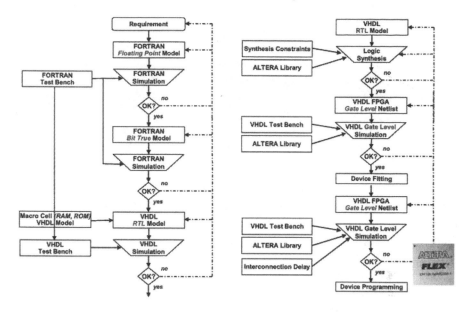

Figure 4-2. MUSIC Receiver FPGA Design Flow.

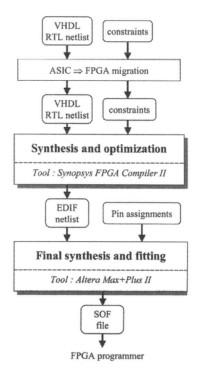

Figure 4-3. FPGA re-targeting of the ASIC design flow.

The conclusion is that when designing for an ASIC implementation the best approach is to include test and other technology specific structures from the very beginning (see Chapter 5 for details). When developing an RTL code no different approaches are needed for ASIC and/or FPGA, except for possible partitioning of the whole circuit into multiple FPGAs. The best approach is thus using a compatible synthesis tool, so that (in principle) the same code can be re-used to produce the same functionality. Developing a unique code for the two designs helps increasing the reliability of the proto-type.

Of course, technology specific macro cells, such as RAM/ROM, micro (DSP) cores, PLLs, physical interfaces, I/Os, clock buffers, cannot be di-rectly ported from one technology to the other, and they need manual re-mapping. Technology specific macro cells can be classified into two catego-ries: cells that can be implemented/modeled in FPGA technology and cells that cannot. When migrating from ASIC to FPGA design, macro cells that cannot be mapped directly into the FPGA (for instance, an ASIC DSP core) need to be implemented directly on the board using off the shelf compo-nents, test chips, or other equivalent circuits. So when developing the HDL code it is good practice to place such macrocells into the top level, so as to minimize and 'localize' the changes that are needed when retargeting to FPGA. This approach also facilitates the use of CAD tools. In fact, by prop-erly using the synthesis directives available within the tool, the same HDL code can be actually used for the two technologies. The CAD recognizes those macrocells that can/cannot be synthesized and acts according to the specified technology.

Even macros which can be implemented in FPGA technology need a lim-ited amount of manual re-mapping. The recommended way of doing this re-mapping is instantiating the ASIC macro where it is needed, and then creat-ing another level of hierarchy for instantiating the FPGA macro(s) under-neath. Doing mapping this way allows one to re-use exactly the same code for both designs. The EC-BAID falls in the latter case, since its ASIC design includes only memory macros (see Section 2.2.1 for further details).

Obviously these considerations do not apply to the multi-rate front end or to the synchronization loops, whose design was only targeted to implementa-tion with programmable devices.

2.1   FPGA Partitioning

Implementation of the MUSIC receiver was based on the DSP breadboard called *PROTEO*, provided by STMicroelectronics [stm]. As detailed in Chapter 6, the PROTEO breadboard is based on a pair of Altera FPGAs 10K100A CPLDs (*Complex Programmable Logic Device*), for a full integration capability of 200 Kgates. The breadboard is also equipped with a fixed point 16 bit DSP processor (66 MIPs, ST18952 by STMicroelcctronics) and includes other features like high speed ADC converters and programmable clock generators. The CPLDs were used for real time signal processing functrions, while the DSP was dedicated to (low speed) housekeeping and measurement tasks.

The overall receiver complexity exceeds the capability of a single breadboard; on the other hand, the final goal of the project was an ASIC implementation of the EC-BAID circuit. Therefore, it was straightforward to resort to *two* identical breadboards. The first PROTEO was dedicated to the implementation of the whole receiver with the exception of the EC-BAID, and was arranged so as to inter-operate either with a second, identical PROTEO connected through a flat cable and implementing the EC-BAID functions, or directly with the ASIC circuit. This solution allowed rapid prototyping of the EC-BAID circuit on FPGA first, and thorough testing of the EC-BAID ASIC subsequently. Figure 4-4 shows the final partitioning of the receiver, implemented either with FPGA+FPGA or with FPGA+ASIC technology. BER measurement and SNIR estimation, as well as housekeeping and initialization, are carried out by the DSP. Figures 4-5 and 4-6 show the schematic and the appearance, respectively, of the FPGA+FPGA configuration.

Preliminary synthesis runs showed that owing to the limited size (100 kGates) of each CPLD mounted on the PROTEO breadboard a single device implementation of the MUSIC receiver front end and EC-BAID detector was not feasible. Hence, further partitioning of both circuits (the front end on PROTEO I, and the EC-BAID on PROTEO II) between the two available CPLDs was taken into account.

Since the time of this design FPGA technology has improved very much. At the time of writing (mid 2003) FPGAs allowing the integration of more than 1 million equivalent gates are easily available on the market, so that the whole MUSIC receiver would surely fit into a single FPGA device. Anyhow, the experience of system partitioning which we describe here is still valid when thinking of a more complex and integrated multimedia terminal with audio/video codecs and complex channel coding/decoding, according to the general trend shown in Figure 1-11 for technology and algorithm complexity.

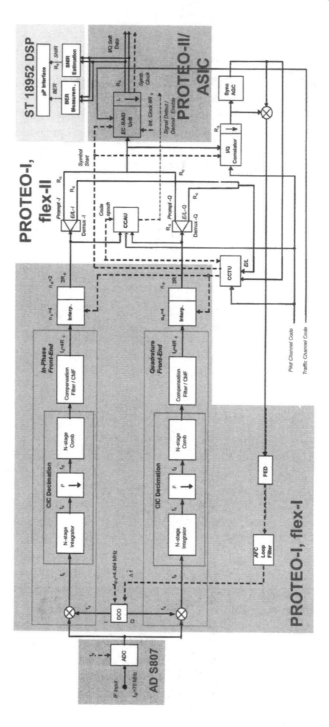

Figure 4-4. MUSIC Receiver Final Partitioning.

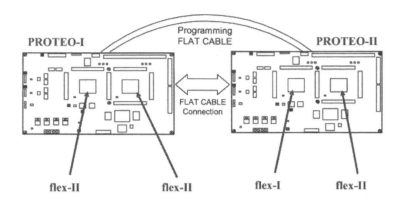

Figure 4-5. Schematic of the FPGA+FPGA based configuration of the receiver.

Figure 4-6. Picture of the two interconnected PROTEO breadboards.

2.1.1   Multi-Rate Front End and Synchronization Circuits on PROTEO-I

Partitioning of the front end and synchronization circuits was quite critical for their final implementation, and heavily affected the resulting architecture. The final mapping of functions onto the available devices was derived
as a trade off between hardware complexity and functional behavior. According to Figure 4-5, the first device on the left hand side of the PROTEO-I
platform, connected to the tri-states buffers that manage the incoming signal,
will be referred to as flex-I, whilst the second one will consequently be
referred to as flex-II. Flex-I was dedicated to front end functionalities: digital
downconversion to baseband, decimation by means of the CIC decimator
filter, chip matched filtering and linear interpolation. The AFC was also
placed here, in order to keep the 11 bit frequency error control signal as
close as possible to the DCO. The AFC is fed with the pilot symbols coming
from the despreader located on flex-II that are already amplitude regulated.
This prevents the use of additional despreading/amplitude correction stages
into flex-I, whose complexity figure, as is shown in the following, is crucial.
Transmission of the pilot symbol values required the implementation of a
simple serial transmission protocol between the two boards, via *Parallel
In/Serial Out* (PISO) and *Serial In/Parallel Out* (SIPO) modules located on
flex-II and flex-I, respectively. This was done to keep the number of connected I/O pins as small as possible.

Similar considerations apply to `fract_del`, the signal which is generated by the CCTU on flex-II and brings back to the linear interpolator located on flex-I the information about the re-sampling epoch. However,
because of its small size (5 bits) a straightforward parallel transmission was
possible. Flex-II takes in the sample stream at rate $4R_c$ (four samples per
chip interval) output by the linear (I/Q) interpolators on flex-I, and is mainly
designated to carry out coarse code acquisition by means of CTAU, and chip
timing tracking performed by the CCTU. As described in Chapter 3, the
CCTU and AFCU are equipped with the SAC unit, which complements the
coarse automatic control loop (AGC) on the IF analog board so as to keep
the signal level constant and independent of the signal to noise plus interference ratio (SNIR). Once again, owing to restrictions on the I/O pins budget
of flex-I, the connectivity functions towards the EC-BAID circuit were
allocated to flex-II: PROTEO-II during rapid prototyping, and the plug-in
mini-board with the ASIC circuit later, were mounted on the upper left 40
pins connector of flex-II. The output of flex-II is the stream of the on time
samples at rate R_c (chip rate) coming from the interpolators on flex-I, after
subsequent decimation. Figures 4-7 and 4-8 outline the partitioning described above between the two CPLDs and show the relative pin-out. Some

of the internal signals are highlighted, together with their respective word length. The main configuration parameters such as rho_CIC, elle_cod, etc., are also reported.

Finally, both devices are connected to the DSP processor via a dedicated, embedded bus, and a suitable interface unit (not shown in Figure 4-7 and 4-8). Communications with the DSP is required at start up to set the proper receiver configuration, as well as to allow for internal receiver monitoring. These features will be described in more detail in Chapter 6.

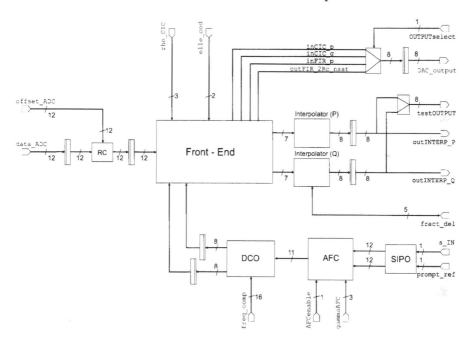

Figure 4-7. PROTEO-I partitioning: flex-I device.

2.1.2 EC-BAID on PROTEO-II

The bottleneck of the FPGA implementation of the EC-BAID was found to be the memory macrocells. The Altera Flex10K100A FPGA contains 12 *Embedded Array Blocks* (EAB), allowing the implementation of RAM blocks of 2048 bits each. Since the EC-BAID circuit needs two different RAMs cuts (128×43 and 384×46 bits) for a total size of 23168 bits, a single CPLD seemed to be enough. Unfortunately the Max+Plus II™ (release 11) synthesis tool can not perform optimized memory cuts whose size is not a power of 2, so it was necessary to use both devices on PROTEO-II, splitting the EC-BAID circuit in two units, named EC-BAID_1 and EC-BAID_2 henceforth, each one managing one memory block.

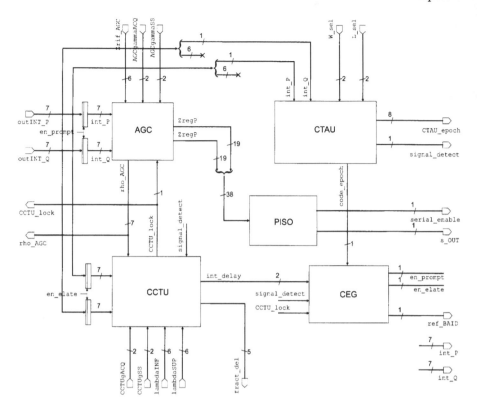

Figure 4-8. PROTEO-I partitioning: flex-II device.

The original VHDL unit was therefore partitioned into two sub-designs, keeping as low as possible the number of signal buses between them. Nevertheless, the 41 bit embedded bus connecting the two FPGAs could not carry all the signals, so that an additional flat cable was needed between the two FPGAs on the EC-BAID PROTEO-II board.

The PROTEO-II partitioning is shown in Figure 4-9, whilst Figure 4-10 shows the arrangement of the two PROTEO breadboards (MUSIC receiver and EC-BAID) corresponding to Figure 4-6, and the flat cable buses between the four FPGAs. In particular, Figure 4-9 shows the VHDL blocks partitioning into the two FPGAs, along with the main internal signals. A detailed description of the EC-BAID architecture will be outlined in Chapter 5, where its ASIC implementation is extensively addressed. In short, block *1* in Figure 4-9 is the standard correlator receiver (CR), while the additional adaptive interference mitigation component is computed in the correlator *2*; after the sum, the output level is regulated by the local AGC (block *3*). RAM *4* stores the last 3·L input samples, which are selected in turn (block 5) to perform hardware multiplexing; the adaptive coefficient vector **x** is stored in RAM *6* and is made orthogonal to the code sequence by block 7. The CPRU (block 8) performs rotation of either EC-BAID or the CR outputs; at the output of the CPRU, multiplexer *9* routes the selected bus into the secondary (auxiliary) output, and multiplexer *10* selects the desired output to be sent back to the MUSIC breadboard for monitoring and testing. Hence multiplexers *9* and *10* are the only blocks added in the FPGA implementation to increase circuit observability.

As mentioned above, partitioning mainly aimed here at splitting the two RAM blocks on two separate devices. Accordingly, RAM *6* was integrated into flex-I of PROTEO-II along with its surrounding logic, whilst the rest of the circuit was assigned to flex-II of PROTEO-II. The other partitioning requirement was to keep the number of the I/O signals as low as possible to comply with the breadboard layout. This is why multiplexers *9* and *10* are needed to select only one output bus at a time. All buses between the two breadboards are carried by a flat cable, which interconnects the MUSIC receiver (flex-II in PROTEO-I) to the EC-BAID (flex-II in PROTEO-II), and they are summarized in Table 4-1. The two FPGAs in PROTEO-II are connected by the internal embedded board bus and by an additional flat cable as summarized in Table 4-2.

2.2   Implementation Details

The MUSIC receiver was described hierarchically in a VHDL mixed structural and behavioral style. Most circuit data paths were developed through a structural description made up of high level arithmetic operators (such as adders, comparators and so on) mixed with explicit lower level instantiation of registers or memories. A behavioral functional description based on VHDL processes was reserved for finite state machines and for control units.

The design is fully synchronous in order to simplify the logic synthesis through utilization of a unique clock tree. However, several clock rates are requested to tick the different building blocks of the receiver. Appropriate multi-rate timing signals were derived from the master clock of the board with the aid of enable strobes that selects only a subset of the active edges of the main clock. In the FPGA+FPGA configuration, PROTEO-II is actually slaved to PROTEO-I. The master clock of the latter comes to the former through the coaxial cable sketched in Figure 4-10. This ensure synchronous operation of the two boards, *provided that* the propagation delay of the coaxial cable is compensated for by (manual) fine tuning of the master clock skew controller on PROTEO-II.

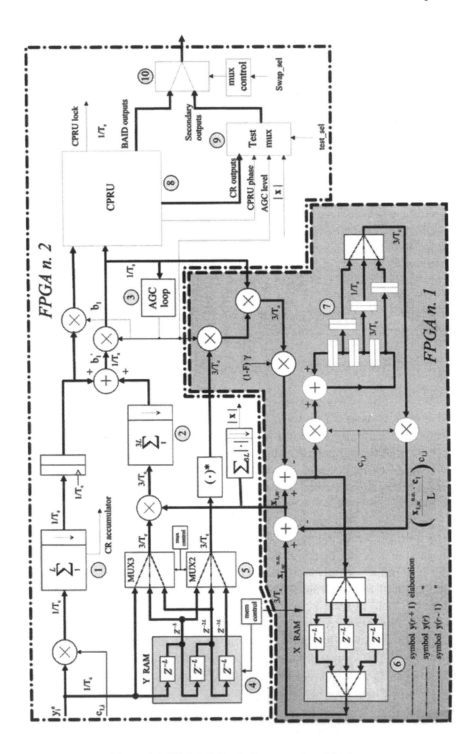

Figure 4-9. EC-BAID block diagram and partitioning.

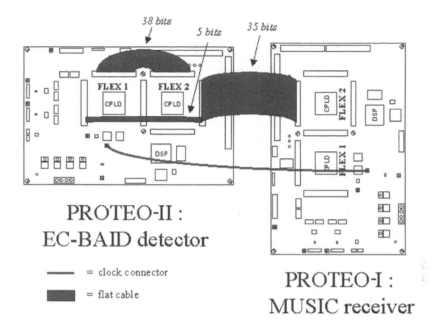

Figure 4-10. Programmable breadboards layout and naming conventions.

Table 4-1. I/O signals of the two PROTEO DSP boards.

I/O signals	Bits	direction
Front end outputs	7 + 7	PROTEO-I ⇒ PROTEO-II
Control signals	5	PROTEO-I ⇒ PROTEO-II
EC-BAID outputs	4 + 4	PROTEO-II ⇒ PROTEO-I
Control signals	3	PROTEO-II ⇒ PROTEO-I

Table 4-2. I/O signals between flex-I and flex-II in PROTEO-II.

I/O signals	bits	direction	Notes
X adaptive vector	10 + 10	flex-I ⇒ flex-II	complex signal
Detector outputs	8 + 8	flex-II ⇒ flex-I	complex signal
Detector inputs	7 + 7	flex-II ⇒ flex-I	complex signal
Code sequence chip	1	flex-II ⇒ flex-I	
AGC level	8	flex-II ⇒ flex-I	
X RAM addresses	9	flex-II ⇒ flex-I	
Config. parameters	9	flex-II ⇒ flex-I	Code length, adaptation step, etc.
Control signals	6	flex-II ⇒ flex-I	Enables, reset, etc.

2.2.1   Register Transfer Level Description

Since we have pushed our partitioning as close as possible to the complexity constraints on the devices, timing constraints played a key role in the synthesis design flow. As a result our first RTL description largely missed those constraints in a few sections of the receiver. So, as is often the case, iterations between different design flow steps were necessary. In the following, we summarize the main actions that were taken to satisfy our design goals.

MUSIC Receiver Front end

As is customary practice, possible timing violations in logic paths required the insertion into the architecture of the receiver front end of several pipelining registers. This is done to break out long combinatorial paths and thus to keep the combinatorial delays within the due timing constraints. The addition of properly designed registers indeed keeps the transmission delays smaller than the worst case clock interval, but also contributes to increasing system latency, since implementation of a certain function ends up with requiring more (short) clock In spite of this, the front end latency was found to be suficiently small, so that the impact on the receiver behavior was insignificant. In this respect, particular care was devoted to the front end sections of the receiver that process the ADC outputs in real time at the highest rate, namely at the clock rate $f_s = 16.384$ MHz, and to those operating at $f_d = 4R_c$ (four samples per chip). Specifically, the CIC decimator filter, as well as the CMF and Equalizer FIR filter, were provided with extra internal registers so as to make them operate in pipelining.

Also, each input and output port was supported by registers in order to get rid of the delays introduced by the propagation time of the I/O pads. Similarly, RAMs and ROMs were specifically described as fully synchronous blocks: data and address buses were supported by registers, not to add their access time to circuit data paths. In particular, two 256×7 ROM modules were implemented on flex-I to store the first-quadrant quantized samples of the sine function in the DCO. The ROM address is the phase signal, represented by 8 bits (equivalent to 10 bit resolution when considering the four-quadrant extended signal), and the value is the sine amplitude represented by 7 bits (equivalent to 8 bit resolution when considering the sign). A 256×23 bit RAM block was reserved on flex-II to store the averaged parallel correlations in the CTAU.

As the reader may have already observed, the diverse step sizes of the sync loops were set for simplicity to values equal to a power of two. This eases programmability with a simple implementation. In some cases the step sizes are switched from an initial larger value to be used for initial acquisition to a smaller *steady state* (SS) value, yielding optimized performance (see Chapter 3). Table 4-3 shows the mapping rule between floating point values and the relative bit true coded quantities for the step sizes of the front end section.

Just like step sizes, all receiver parameters such as code length, CIC decimation factor, etc., were coded onto with a proper number of bits. Table 4-4 reports the relevant associations.

Table 4-3. Bit true coding of the loops step-sizes.

Step size	Floating point values	Bit True Coding
γ_{CCTU}	$2^{-9}, 2^{-8}, 2^{-7}, 2^{-6}$	'00', '01', '10', '11'
γ_{AGC}	$2^{-5}, 2^{-4}, 2^{-3}, 2^{-2}$	'00', '01', '10', '11'
γ_{AFC}	$2^{-19}, 2^{-18}, 2^{-17}, 2^{-16}, 2^{-15}$	'000', '001', '010', '011', '100'

Table 4-4. Receiver parameters mapping rule.

Parameter	Floating point values	Bit True Coding	Description
L	32, 64, 128	'00', '01', '11'	Code length
ρ_{CIC}	2, 4, 8, 16, 32	'000', '001', '010', '011', '100'	Decimation factor
W	128, 256, 512, 1024	'00', '01', '10', '11'	CTAU smoothing window
λ	1.00, 1.25, 1.50, 1.75	'00', '01', '10', '11'	CTAU threshold

EC-BAID

The interference mitigating detector was implemented in FPGA for rapid prototyping and verification, but it was described in RTL with the final target of ASIC implementation. Therefore no change in the RTL description of the circuit (which will be detailed in Chapter 5) was implemented when it was migrated to FPGA For instance, as opposed to the multi-rate front end design, no pipelining was introduced in the EC-BAID architecture to speed up its clock frequency, or equivalently, the maximum data rate it could process. This motivates the clock speed downgrading of the FPGA EC-BAID implementation that we have already mentioned.

Special attention was also given to the arrangement of the ASIC/FPGA pinout. The flat cable connection between PROTEO-I (receiver breadboard) to PROTEO-II (EC-BAID on FPGA) was designed to be re-used pin by pin when the ASIC EC-BAID implementation take the place of PROTEO-II. The relevant cable pin assignments are listed in Table 4-5.

2.2.2   Logic Synthesis Results

As mentioned above, and according to the design flow of Figures 4-2 and 4-3, the receiver VHDL synthesis was performed by jointly using the CAD tools FPGA Compiler II^{TM} by Synopsys, and MAX+Plus II^{TM} by Altera.

Starting from the VHDL netlist of the receiver front end at the RTL level, further VHDL blocks were added in order to increase the circuit testability, and a suited timing constraints script file was set up. The main synthesis and optimization effort was carried out by the Synopsys FPGA Compiler II, whose output database was transferred via *Electronic Database Interchange Format* (EDIF) to the Altera MAX+Plus II^{TM} tool where it was utilized as a starting point for the final pad assignment and fitting phases. Resources occupation after the fitting for the MUSIC front end (PROTEO-I) is summarized in Table 4-6 in terms of Logic Elements (LEs) and Embedded Array Blocks (EABs).

Table 4-5. Flat cable connection between PROTEO-I and PROTEO-II/ASIC board.

Connector pin[*]	Signal name[**]	Direction[***]
40..37	Unused	-
36	Lock	Input
35	Test_se	Output
34	Test_si	Output
33	Tm	Output
32	Bact	Output
31	Rack	Output
30	Txt	Output
29	Resn	Output
28	Sym_in	Output
27	Enc8	Output
26..20	Yr[6..0]	Output
19..13	Yi[6..0]	Output
12	Req	Input
11	Sym_out	Input
10..7	Outr[3..0]	Input
6..3	Outi[3..0]	Input
2,1	Reserved (3.3 V power supply and ground)	Input

[*]   *MUSIC receiver breadboard J8 pins*
[**]   *ASIC I/O names as described in Chapter 5*
[***] *Direction with respect to the MUSIC receiver breadboard side*

Table 4-6. Fitting results of PROTEO-I breadboard.

	LE	EAB
flex-I	91 %	3072 bits
flex-II	88 %	5888 bits

Tables 4-7 and 4-8 report the breakdown of the synthesis results of flex-I and flex-II on PROTEO-I, respectively. For each building block the required number of LEs and EABs is reported, and the utilization factor is computed. For comparison we recall that 4992 LEs and 12 EABs are overall available on each device. Finally, the resource occupation factor of the EC-BAID after fitting on PROTEO-II is summarized in Table 4-9, while the relevant synthesis breakdown of flex-I and flex-II is reported in Tables 4-10 and 4-11, respectevely.

Table 4-7. Synthesis breakdown of flex-I – PROTEO-I.

	LE	EAB
Front end	3241	0
Interp	433	0
DCO	40	2
AFC	640	0
Master	66	0
DSP/IF_1	154	0
Total cell count	4574	2
Utilization	91%	14%

Table 4-8. Synthesis breakdown of flex-II – PROTEO-I.

2nd CPLD	LE	EAB
CTAU	1262	3
CCTU	1546	0
Debug	310	0
AGC	1018	0
DSP/IF_2	289	0
Total cell count	4425	3
Utilization	88%	23%

Table 4-9. Fitting results of PROTEO-II breadboard.

	LE	EAB
flex-I	54 %	23552 bits
flex-II	67 %	5504 bits

Table 4-10. Synthesis breakdown of flex-I – PROTEO-II.

	LE	EAB
EC-BAID_1	2728	12
Total cell count	4992	12
Utilization	54%	95%

Table 4-11. Synthesis breakdown of flex-II – PROTEO-II.

	LE	EAB
EC-BAID_2	3370	4
Total cell count	4992	12
Utilization	67%	22%

The receiver architecture was successively refined until the static timing analysis, performed with the conservative EPF10K100 ARC240-3 timing model, showed no timing violations. After that the receiver was finally able to operate at specified system chip rate as in Table 2-3.

The same analysis, when repeated for the FPGA implementation of the EC-BAID, showed failure at R_c = 1024 kchips/s. In particular, the Altera Static Timing Analyzer located the critical path on the bus Bin, connecting flex-II to flex-I. Here the timing requirement is $T_c/4$, and the signal may take up to 300 ns to go from register to register. Because of that, the higher testable chip rate turned out to be R_c = 512 kchip/s ($T_c/4$ = 488 ns). Fortunately, the timing values reported by the static timing analysis are based on a conservative timing model of the particular Flex10K100A version, namely the EPF10K100 ARC240-3. This is why the hardware testing steps (described in Chapter 6) were in the practice successfully completed even at the chip rate R_c = 1024 kchip/s, although the theoretical analysis based on Altera Static Timing Analyzer had failed. Anyway, to be on the safe side, and bearing in mind that the goal of this FPGA implementation step was just a functional validation of the circuit design, no BER measurements were performed on the FPGA+FPGA configuration at a chip rate greater than 512 kchip/s.

Chapter 5

INTERFERENCE MITIGATION PROCESSOR ASIC'S DESIGN

Is it difficult to design a CDMA receiver mitigating interference? It is certainly challenging, but it is no more difficult than designing a conventional DS/SS receiver with some additional intelligence and processing power. The previous Chapters have shown the 'conventional' side of the design. This Chapter, on the contrary, is focused on the value-added core of the MUSIC receiver: the details of the ASIC design for the interference mitigation processor, the so called EC-BAID. Starting with a description of the ASIC I/O interface (with details on the circuit pin-out along and on the timing diagram of the input/output signals) the chapter develops through to an overview of the serial protocol which is used for the configuration of the ASIC, followed by an overall portrayal of the circuit and by detailed descriptions of each sub-block. Finally, the Front to Back ASIC design flow is presented together with the resulting circuit statistics for a 0.18 μm CMOS technology implementation.

1.   ASIC INPUT/OUTPUT INTERFACE

Definition of the I/O interface is one of the major drivers in the ASIC design cycle and must be considered since the very beginning of the process. The preliminary feasibility study told us that the EC-BAID circuit is characterized by a small gate complexity, which implies a small ASIC core area and a pad limited layout in the selected technology (HCMOS8D by STMicroelectronics, see Section 3.1). For this reason, in order to reduce the size of the circuit the number of I/O pins was kept as low as possible, and a 44 pin package was selected. The limitations caused by such choice in the receiver interface were dealt with by proper output multiplexing, and by serially loading all the EC-BAID configuration parameters at startup.

1.1   ASIC Pin-Out

The pin-out of the EC-BAID ASIC is shown in Figure 5-1, while a short description of each pin function is presented in Table 1. The selected 44 pin package is the TQFP44, which bears an external side length of 10 mm. Two power supplies are required, as the core circuit works at 1.8 V while the I/O pads must support a power supply of 3.3 V, in order to correctly operate with the signals of the MUSIC receiver board.

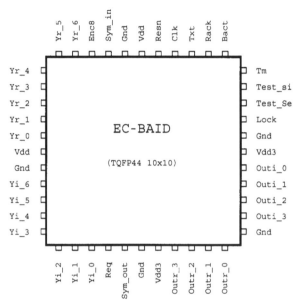

Figure 5-1. EC-BAID ASIC pin-out.

The EC-BAID circuit makes use of fully synchronous logic, requiring a single external clock input from the MUSIC receiver breadboard (the Clk pin), while different internal sampling rates are implemented by means of appropriate enable signals. All of the internal registers sample their inputs on the positive edge of Clk, provided that the corresponding enable strobe is high. As explained in Section 3.2, the circuit was synthesized to work at the clock frequency of 32.768 MHz with a wide margin (the actual timing constraints during the synthesis were placed at 40 MHz), with the goal of a maximum chip rate of 4.096 MHz. However, according to the MUSIC specifications (see Chapter 1), the receiver breadboard drives the EC-BAID ASIC with a clock frequency f_s = 16.384 MHz and processes signals with chip rates ranging from 0.128 to 2.048 Mchip/s.

The Enc8 input is an external synchronization signal which enables a clock rising edge every $T_c/8$ seconds, where $T_c=1/R_c$ is the chip period. The clock is enabled if Enc8 is high. The need for an operating rate eight times higher than the chip rate arises from the hardware multiplexing feature (actually, internal arithmetical operations are performed at rate $4 \cdot R_c$) together with synchronous SRAM utilization whereby one read cycle and one write cycle occur every $T_c/4$ seconds. As a consequence the maximum allowed chip rate is $R_{c,\max} = f_s/8$ (e.g., 2.048 Mchip/s @ $f_s = 16.384$ MHz).

According to the Chip Clock Tracking Unit algorithm (CCTU, described in Chapter 3) sometimes the time reference of a CDMA symbol is delayed or anticipated by $T_c/4$ to track the transmitted chip clock. By assuming the EC-BAID frequency clock 8 times faster than the chip clock frequency, a proper sampling of the input samples with no lost of data is guaranteed. This is true even in the presence of a shorter symbol period, when in response to the CCTU algorithm, the last chip of the sequence only lasts $3T_c/4$ instead of the nominal T_c. As shown in Section 2.1.8, the EC-BAID can operate in each of these scenarios (symbol realignment of $-T_c/4$, 0 or $T_c/4$). Whenever an enable pulse is present on the symbol start reference Sym_in the circuit starts sampling and processing L input chips (where L is the code repetition period). If no more enable strobes are coming, the circuit stops its internal operations, waiting to resume at the next Sym_in pulse.

Table 5-1. EC-BAID ASIC pins description.

Pin number	Signal Name	Direction	Description
43,44,1–5	Yr[6:0]	Input	EC-BAID input signal, in phase (chip rate)
8–14	Yi[6:0]	Input	EC-BAID input signal, quadrature (chip rate)
15	Req	Output	Parameters transmission request
16	Sym_out	Output	Output symbol reference
19–22	Outr[3:0]	Output	Configurable output, phase (symbol rate)
24–27	Outi[3:0]	Output	Configurable output, quadrature (symbol rate)
30	Lock	Output	CPRU lock indicator (1 = locked)
31	Test_se	Input	Test scan enable
32	Test_si	Input	Test scan input
33	Tm	Input	Test mode (0 = normal op., 1 = test mode)
34	Bact	Input	BIST activation (1 = start of BIST procedure)
35	Rack	Input	Parameters transmission request acknowledgment
36	Txt	Input	Parameters transmission bit
37	Clk	Input	System clock
38	Resn	Input	Synchronous reset, active low
41	Sym_in	Input	Input symbol reference
42	Enc8	Input	Clock enable at rate $T_c/8$
18,28	Vdd3		3.3 V power supply
6,39	Vdd		1.8 V power supply
7,17,23,29,40	Gnd		Ground

The timing diagram of the ASIC RTL behavioral simulation, shown in Figure 5-2, illustrates the input sampling operations. First, the synchronization signals Sym_in and Enc8 (shown in Figure 5-2 with the internal VHDL names Symbref_unreg and Enc8_unreg) are buffered to prevent exceedingly long combinatorial paths between the MUSIC receiver and the EC-BAID registers and outputs.

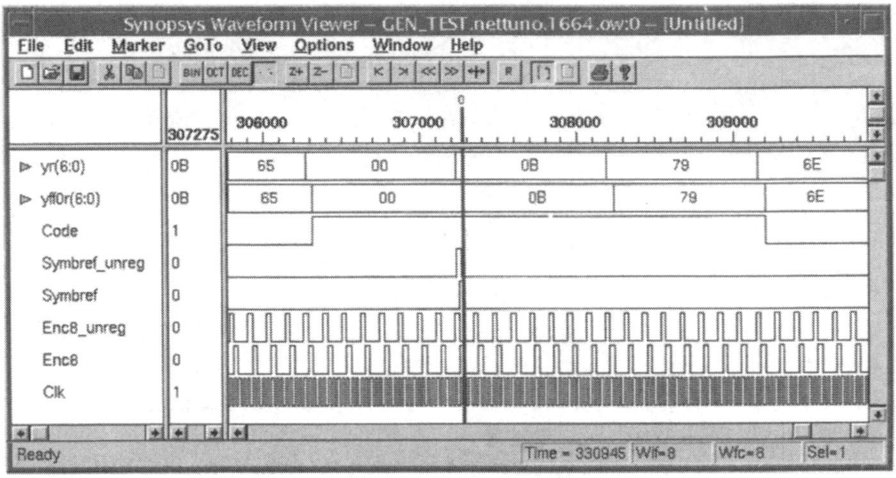

Figure 5-2. Input sampling related signals.

Therefore all the sampling operations are enabled by these delayed replicas of the strobe signals (denoted with the VHDL names Symbref and Enc8). As an example, the clock edge highlighted in Figure 5-2 is enabled by the delayed Enc8 strobe and it triggers sampling of the input signal Yr[6:0] in a register which drives the Yff0r[6:0] bus[1].

The 44 pin package entails some limitations on the bus width of the I/O signals, so that, in order to keep the ASIC pin number low, all the desired output signals are multiplexed into a single configurable 8 bit wide bus. This bus is made up by the Outr[3:0] and the Outi[3:0] outputs, where Outr[3] is the Most Significant Bit (MSB) and Outi[0] is the Least Significant Bit (LSB). The main ASIC output signals are the symbol rate signal strobes at the despreader output coming from the EC-BAID receiver (with VHDL names Boutr[3:0] and Bouti[3:0]). Also, an auxiliary output (Auxr[3:0] plus Auxi[3:0]) is driven by a multiplexer which can select among four further signals according to the out_sel configuration parameter (see Table 5-2). The ASIC outputs meaning is then controlled by the swap_sel parameter (see Table 5-3): if swap_sel is set to 0 the EC-BAID outputs only (Boutr and Bouti) are sent out, while setting it to 1, will cause both the EC-BAID and auxiliary outputs (Boutr, Bouti and

---

[1] The pin names Yr_6 ... Yr_0 of the ASIC correspond to the internal Yr[6:0] bus, and a similar convention is used for the Yi[6:0], Outr[3:0] and Outi[3:0] buses.

Auxr, Auxi) to be multiplexed together, half a symbol period each, as in the example shown in Figure 5-3.

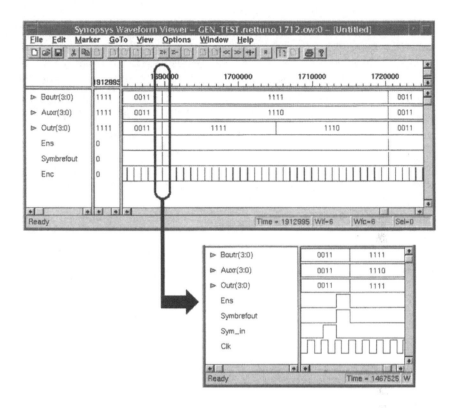

Figure 5-3. Output selection and synchronization.

Figure 5-3 also shows Sym_out signal generation (with the internal VHDL name Symbrefout). This reference output signal is high on the same clock edge where the outputs are buffered, and therefore it is aligned with the internal symbol reference strobes (Ens, Symbref) which in turn are delayed with respect to the external input reference Sym_in, as previously explained. The reset and initialization operations start when the Resn input goes to zero. This external reset is buffered in a three flip flop chain in order to reduce metastability effects. The resulting signal is used as a synchronous, active-low reset for most of the internal registers. When Resn is sampled at a low value the whole circuit is stopped, whilst when the reset is released two operations are performed before starting normal processing: first, the configuration parameters are serially loaded together with the code sequence bits, then internal RAMs are loaded with zero values (and this operation takes one more symbol period). This initialization procedure is sketched in Figure 5-4[2]. Once initialization is accomplished, the EC-BAID circuit is ready to process the input chips. Possible Sym_in pulses sent before the end of these phases are ignored.

As a further method to reduce the I/O pins number, all the configuration parameters, including the code chip sequence, are serially loaded through the Req, Rack and Txt signals. The simple handshake protocol shown in Figure 5-5 is initiated by the ASIC when it sets the Req signal high. The MUSIC receiver breadboard then sends an information bit through the Txt pin and concurrently sets the Rack signal high to instruct the EC-BAID to read the Txt bit. Finally, the ASIC sets the Req bit low and waits for a low value on the Rack pin in order to complete the handshake. The whole procedure is repeated for a total of $L + 57$ bits: the 2 bit representation of the code length L first, followed by the L binary chips of the user code sequence (to be saved into a column of the RAM), and ending up with 55 more configuration bits to be stored in a shift register. More details about the order and the meaning of the various parameters are given in the next subsection.

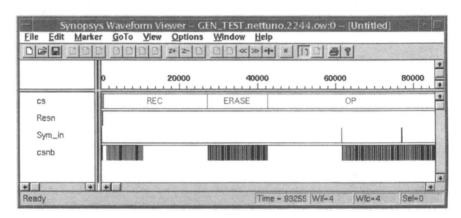

Figure 5-4. Initialization phases.

Table 5-2. Auxiliary output selection.

Out_sel[1:0]	Auxr[3:0] and Auxi[3:0] auxiliary outputs (8 bits)
00	Outputs of the standard correlation receiver (4 + 4 bits)
01	Carrier phase estimated by the CPRU (8 bits)
10	Internal AGC gain level (8 bits)
11	Modulus of the EC-BAID x^e adaptive vector (8 bits)

---

[2] The csnb waveform in Figure 5-4 is a RAM enable signal whilst cs is the current state of the main synchronization block.

Table 5-3. ASIC outputs configuration.

swap_sel	Outr[3:0] and Outi[3:0] ASIC outputs
0	Boutr[3:0] and Bouti[3:0] for all the symbol period
1	Boutr[3:0] and Bouti[3:0] in the first symbol semi-period, Auxr[3:0] and Auxi[3:0] in the second semi-period.

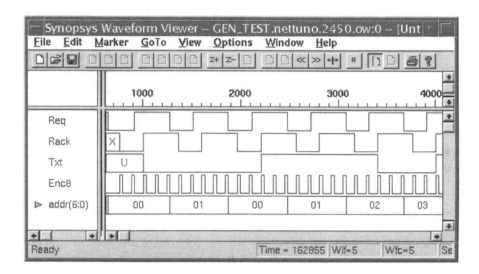

Figure 5-5. Configuration parameters loading.

1.2 Configuration Parameters

The whole configuration sequence is summarized in Table 5-4, where bit number 0 represents the first bit received by the ASIC. After the code length and the whole code sequence bits, various parameters which allow us to configure the ASIC functionality and to specify the values for the algorithm parameters are exchanged. Brif and agcgamma refers to the AGC loop which is detailed in Section 2. Winlen and wintype define the window length of the EC-BAID correlation as follows: with wintype equal to 0, $3L$ input chips (L is the code length) are processed for the detection of each information symbol, while with wintype equal to 1 the correlation is computed on an L-chip symbol interval plus only a portion (whose width, in chips, is specified by winlen) of the previous and the next symbol intervals, yielding a total window length of $L + 2 \cdot$ winlen chips. Costason-off is the CPRU enable bit, whilst gammacostas and rhocostas are the adaptation steps of the CPRU second order loop, respectively. The parameters involved in the phase lock detector are Lock (the adaptation step) and threshigh, threslow (the threshold values of the lock detecting circuit). The bit ec12sel selects the desired EC-BAID algorithm version (see Chapter 3) as follows: if it is set to 1, the 'chunk' orthogonality condition (3.110) is imposed on the adaptive vector $\mathbf{x}^e$ (where the superscript ^{e} stands for extended, i.e., $3L$ elements long), while setting it to 0, causes the orthogonality constraint to be imposed only to the central part $\mathbf{x}_0$ of $\mathbf{x}^e$ (see (3.115)). Leakenable is the configuration bit enabling a 'leakage' correction to the EC-BAID algorithm, as detailed in Section 2.1.4, whereby the relevant factor is selected by the Leak parameter. Finally, Gam encodes the EC-BAID algorithm adaptation step, while swap_sel and out_sel set the outputs behavior as previously detailed in Tables 5-2 and 5-3. The values of the different programmable parameters that were used as a baseline in our testing are shown in Table 5-5.

2.   ASIC DETAILED ARCHITECTURE

This Section deals with the description of the EC-BAID bit true implementation at the register transfer level, which has been the starting point of the Front End design flow. In this respect we remark that all the buses shown in the following block diagrams are bit true representations of the relevant floating point signals, as explained in Chapter 4. The bus sizes have been carefully selected by means of extensive simulation runs as a trade off between circuit complexity and final BER performance. In particular, the VHDL description of some critical sub-blocks relies on variable parameters to specify the signal bit width. Such parameters are reported in the following sub-circuits block diagrams, together with their final values selected for the ASIC circuit.

Figure 5-6 shows the top level block diagram of the whole circuit with all main functional blocks. Starting from the Yr/Yi (soft) input chips, the output symbols are built by adding to the standard correlator output a correction term obtained with the adaptive vector $\mathbf{x}^e$. A further block implements the vector adaptation rule, and a SRAM stores the coefficients of $\mathbf{x}^e$. One other SRAM is needed in order to store the code sequence and the most recent $3L$ input chips. The CPRU block performs carrier phase recovery at symbol rate, and passes its outputs to the output control block, which operates as explained in Chapter 3. The main synchronization block provides timing signals for the initialization phase, while two more sub-block are responsible for parameters loading and generation of the internal enable signals.

In the following Sections the RTL architecture of the main EC-BAID blocks is presented. Signal names, reported in italic in the block diagrams, are those used in the VHDL description, with the convention that complex signals are drawn with bold lines and their names (for example, *sig-*

nal_name) correspond to a pair of VHDL vectors having the same name and suffixes r and i for the real and the imaginary parts, respectively, (for example *signal_namer* and *signal_namei*). When a bus width N is shown for a complex signal it means N bits for the real part and N bits for the imaginary part. An equivalent notation is N,N.

Table 5-4. ASIC Configuration parameters.

Bit number	Parameter	Description
1..0	Lsel[1:0]	Code sequence length L $00 \rightarrow 32$ $01 \rightarrow 64$ $11 \rightarrow 128$
2 ⋮	c[0] ⋮	Code sequence bit #0 ⋮
L+1	c[L]	Code sequence bit #L
L+4..L+2	agcgamma[2:0]	AGC adaptation step $\gamma_{AGC} = 2^{(\text{agcgamma-5})}$
L+10..L+5	Brif[5:0]	AGC reference level $b_{REF} = B_{rif} \cdot 2^{-5}$
L+17..L+11	winlen[6:0]	Extended window side lobe length in chips
L+18	wintype	$0 \rightarrow$ Full window length ($W_{len} = 3L$) $1 \rightarrow$ Shortened window length ($W_{len} = L+2 \cdot$ winlen)
L+19	costasonoff	CPRU enable $0 \rightarrow$ CPRU off $1 \rightarrow$ CPRU on
L+20	ec12sel	EC-BAID version $0 \rightarrow \mathbf{c}^T \mathbf{x}_0 = 0$ $1 \rightarrow \mathbf{c}^T \mathbf{x}_w = 0$ with w=-1, 0, 1
L+21	swap_sel	Outr[3:0]/Outi[3:0] outputs control $0 \rightarrow$ EC-BAID (T_s), $1 \rightarrow$ EC-BAID ($T_s/2$) / Auxiliary outputs ($T_s/2$)
L+23..L+22	rhocostas[1:0]	CPRU loop second parameter $\rho_c = 2^{(\text{rhocostas - 10})}$
L+25..L+24	gammacostas[1:0]	CPRU loop first parameter $\gamma_c = 2^{(\text{gammacostas - 10})}$
L+27..L+26	leak[1:0]	Leak factor $F = 2^{-(1+\text{leak})}$
L+28	leakEnable	Leakage enable $0 \rightarrow$ Leakage off $1 \rightarrow$ Leakage on
L+31..L+29	gammalock[2:0]	CPRU lock detector adaptation speed $\gamma_{lock} = 2^{(\text{gammalock - 12})}$
L+41..L+32	threslow[9:0]	CPRU lock detector low threshold $T_{low} =$ threslow$\cdot 2^{-8}$
L+51..L+42	threshigh[9:0]	CPRU lock detector high threshold $T_{high} =$ threshigh$\cdot 2^{-8}$
L+53..L+52	out_sel[1:0]	Auxiliary outputs selection $00 \rightarrow$ Standard correlator $01 \rightarrow$ CPRU carrier phase $10 \rightarrow$ AGC level $11 \rightarrow |\mathbf{x}^e|$ estimation
L+56..L+54	gam[2:0]	EC-BAID algorithm adaptation step $\gamma = 2^{(\text{gam-17})}$

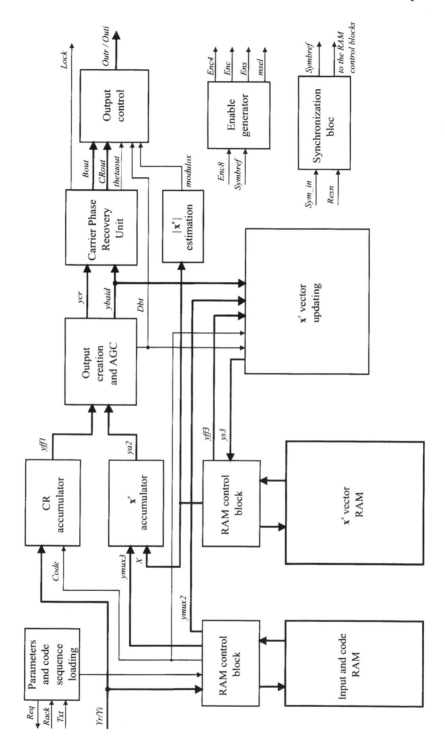

Figure 5-6. ASIC functional block diagram.

Table 5-5. Suggested (default) values of the configuration parameters.

Parameter	Suggested value
Brif	45
rhocostas	1
gammacostas	1
gammalock	1
threslow	461
threshigh	464

The block diagrams in this Section report the bit true RTL descriptions of the EC-BAID sub-blocks, whilst their functionality is illustrated via the usual floating point equations (see Chapter 3). Floating point elements use the same vector notation as in Chapter 3, which is briefly summarized in Table 5-6. The associations between floating point signals and the relevant bit true representations are explained in the following Section.

We conclude this sub-Section with a explicative remarks about notation. As introduced in Chapter 3, we will denote with $\mathbf{y}(r)$ the array containing the L chip rate samples relevant to the rth information symbol. This array is related to the sub-vectors of the extended vector $\mathbf{y}^e(r)$ as follows (see last row in Table 5-6):

$$\mathbf{y}_{-1}(r) = \mathbf{y}(r-1), \quad \mathbf{y}_0(r) = \mathbf{y}(r), \quad \mathbf{y}_1(r) = \mathbf{y}(r+1). \tag{5.1}$$

Also, we denote by $y_i(r)$, c_i, $x_i(r)$, $e_{w,i}(r)$, $x_{w,i}(r)$ and $\Delta x_{w,i}(r)$ the ith components of the vectors $\mathbf{y}(r)$, $\mathbf{c}$, $\mathbf{x}^e(r)$, $\mathbf{e}_w(r)$, $\mathbf{x}_w(r)$ and $\Delta \mathbf{x}_{w,i}(r)$, respectively. Some of these vectors have already been defined in Chapter 3, whilst the others will be introduced later in this Chapter.

Table 5-6. Floating point vector notation.

Notation	Description
$\mathbf{c}$	Code sequence, L elements
$\mathbf{c}^e = [\mathbf{0}, \mathbf{c}^T, \mathbf{0}]^T$	Code sequence extended with zeroes, $3L$ elements
$\mathbf{x}^e = [\mathbf{x}\text{-}_1^T, \mathbf{x}_0^T, \mathbf{x}_1^T]^T$	EC-BAID adaptive vector, $3L$ elements
$\mathbf{x}_w$, with w=-1, 0, 1	EC-BAID adaptive vector sub-blocks, L elements each
$\mathbf{y}(r)$	Array of L input chips in the rth symbol
$\mathbf{y}^e(r) = [\mathbf{y}_{-1}^T, \mathbf{y}_0^T, \mathbf{y}_1^T]^T = [\mathbf{y}(r\text{-}1)^T, \mathbf{y}(r)^T, \mathbf{y}(r\text{+}1)^T]^T$	Array of $3L$ input chips centered on the rth symbol

2.1   Bit True Architecture

All the VHDL bit vectors which appear in the block diagrams in Figure 5-6 are 'bit true' representations of the relevant floating point quantities in the algorithm equations. Each floating point signal, for example the ith component $x_{w,i}$ of the adaptive vector $\mathbf{x}_w$, is represented by an integer value, for example X, with a proportionality relation

$$X = \text{int } \{x_{w,i} \cdot SF\}, \qquad\qquad (5.2)$$

where the scaling factor SF 'centers' the value of the signal within the fixed point representatinn. Typically, the FP range of our signals is ±1, so that the default value of the scaling factor is 2^N when N bits are used for their bit true representation.

In order to reduce the circuit complexity with a minimum impact on the BER performance, some well known design techniques were adopted. For example, bus sizes are kept under control by discarding LSBs where possible, or by saturating signals between proper levels, as sketched in Figure 5-7.

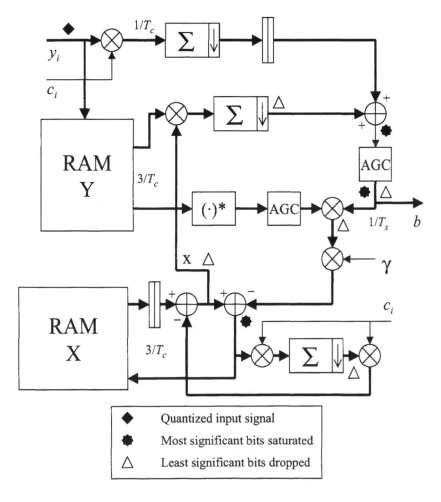

Figure 5-7. Key positions in the bit true dimensioning of signals.

2.1.1 Correlation Receiver

Conventional despreading/correlation is performed by the circuit shown in Figure 5-8, wherein the accumulator A1 sums L input chips within a symbol period (T_s). The internal register of A1 is reset at each symbol start by a control signal (not shown in the diagram). Register FF0 holds the last (soft) chip value, while register FF1 introduces a T_s delay in order to properly synchronize the subsequent operations. A saturation block constraints the input values within the range $\left[-\left(2^{N-1}-1\right),\left(2^{N-1}-1\right)\right]$. The NORM1 block performs left-shift by 7 - $\log_2 L$ bits, so that the final value of $yff1$ follows the relation

$$yff1 \; \propto \; \frac{\mathbf{c}^{e^T}\mathbf{y}^e}{L}. \tag{5.3}$$

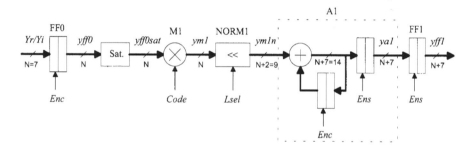

Figure 5-8. Correlation Receiver.

2.1.2 Adaptive Interference Mitigation

The EC-BAID algorithm mitigates the multiple access interference (MAI) by adding an adaptive 'mitigation' vector $\mathbf{x}^e$ to the code sequence $\mathbf{c}^e$ and by computing correlation over a window extended to a maximum of $3L$ chips. The resulting output symbol is then

$$b(r)=\frac{\left[\mathbf{c}^e+\mathbf{x}^e(r)\right]^T \mathbf{y}^e(r)}{L}=\frac{\mathbf{c}^{e^T}\mathbf{y}^e(r)}{L}+\frac{\mathbf{x}^e(r)^T \mathbf{y}^e(r)}{L}, \tag{5.4}$$

where the first term in the rightmost side expression is the conventional correlator output, whilst the second one is the so called 'adaptive correction term', which is obtained from the sub-block of Figure 5-9.

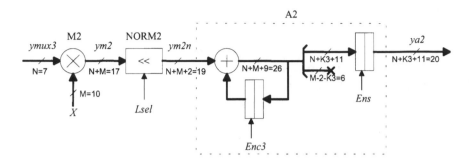

Figure 5-9. MAI adaptive correction term.

The signals X and *Ymux3* are proportional to the floating point values of the elements in $\mathbf{x}^e$ and $\mathbf{y}^e$, respectively. Also the shift block NORM2 has the same function as that in Figure 5-8. Accumulator A2 adds $3L$ terms every symbol, and its internal register is reset at each symbol start by a control signal (not shown in the diagram). The most significant bits of the accumulated value is passed to the subsequent adder S1 (shown in Figure 5-10), and such scaling is needed to make the output of the standard correlator *yff1* and this output *ya2* compatible.

2.1.3   Automatic Gain Control and Output Generation

The (soft) input chips values are delivered to the ASIC by the MUSIC receiver breadboard, where the received analog signal amplitude is adjusted with respect to the input dynamic range of the ADC. Such a level regulation applies to the whole received signal (i.e., useful channel + interferers + noise), whilst the level of the useful channel within the received signal may considerably vary according to the different SNIR configurations. Precision amplitude control is therefore needed. This is why the EC-BAID ASIC embeds a digital AGC to regulate the level of the signal strobe at the detector output. Assume that the nominal input signal $\mathbf{y}^e$ is as in (2.106), with unit amplitude for both the real and imaginary parts of the useful chanel, and with a variable level of noise and interferers. The signal at the output of the analog IF AGC is then

$$\mathbf{y}_v^e = G_{an} \cdot \mathbf{y}^e, \tag{5.5}$$

where G_{an} is optimum for A/D conversion. The goal of the digital AGC is then to produce a variable gain factor G close to the value $1/G_{an}$, in order to restore a unit amplitude $\mathbf{y}_{reg}^e$ signal:

$$\mathbf{y}_{reg}^e = G \cdot \mathbf{y}_v = (G_{an} \cdot G) \cdot \mathbf{y}^e \cong \mathbf{y}^e, \tag{5.6}$$

In the ASIC architecture the gain factor G is applied to the output signals on the M5, M6, M7 multipliers, rather than directly to the inputs, because such an arrangement allows one to keep the width of the input buses as low as possible, thus reducing the size of the input RAM and the complexity of several arithmetical blocks. As depicted in Figure 5-10, the standard correlator output (*ya2*) and the adaptive correction (*yff1*) are added to build up the output of the EC-BAID algorithm. Denoting with b' the floating point output before amplitude regulation, then the output b after the AGC is built according to the following first order loop equations

$$b = G \cdot b', \tag{5.7}$$

$$\varepsilon_G = |b| - b_{REF}, \tag{5.8}$$

$$G(r+1) = G(r) - \gamma_{AGC} \cdot \varepsilon_G, \tag{5.9}$$

where the error signal ε_G is calculated comparing the output amplitude with a reference value (B_{rif} in Figure 5-10). The amplitude of the complex-valued quantity *agcin* (defined as *agcin* = P +jQ) was approximated as follows [Fan02]:

$$B_{mod} = a + d/2 - (a+d)/16, \tag{5.10}$$

with $a = \max\{|P|, |Q|\}$ and $d = \min\{|P|, |Q|\}$. The approximation allowed us to save a considerable amount of area in the ASIC implementation, with an error that never exceeds 11.8% (7% in our particular operating conditions). The adaptation step of the AGC loop is selected among powers of 2 (coded by *agcgamma*) in order to implement the required multiplication via a simple shift operation.

2.1.4 Storing and Upgrading of the Adaptive Vector

As detailed in Chapter 3, the EC-BAID algorithm is a first-order loop that is based on the following equations:

$$b(r) = \frac{\left[\mathbf{c}^e + \mathbf{x}^e(r)\right]^T \mathbf{y}^e(r)}{L}, \tag{5.11}$$

$$\mathbf{x}_w(r+1) = \mathbf{x}_w(r) - \gamma b(r-1) \cdot \left[\mathbf{y}_w^*(r-1) - \frac{\mathbf{y}_w^*(r-1)^T \mathbf{c}}{L} \mathbf{c} \right],$$ (5.12)

$$\mathbf{c}^T \mathbf{x}_w = 0 \ , \ \ \text{with} \ \ w = -1,0,1,$$ (5.13)

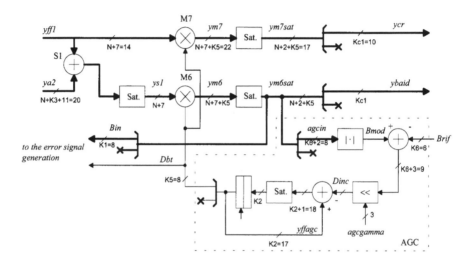

Figure 5-10. AGC and outputs creation.

Typical problems related to the bit true implementation of the loop (as explained in Chapter 3) are prevented by splitting (5.12) in two steps:

$$\mathbf{x}^e(r+1)^{n.o.} = \mathbf{x}^e(r)^{ort.} - \gamma b(r-1)\mathbf{y}^{e^*}(r-1),$$ (5.14)

$$\mathbf{x}_w(r)^{ort.} = \mathbf{x}_w(r)^{ort.} - \Delta\mathbf{x}_w(r) \ , \ \ \text{with} \ \ w = -1,0,1,$$ (5.15)

where

$$\mathbf{x}^e(r)^{ort.} = \left[\mathbf{x}_{-1}(r)^{ort.T}, \mathbf{x}_0(r)^{ort.T}, \mathbf{x}_1(r)^{ort.T} \right]^T,$$ (5.16)

$$\mathbf{x}^e\left(r\right)^{n.o.} = \left[\mathbf{x}_{-1}\left(r\right)^{n.o.^T}, \mathbf{x}_0\left(r\right)^{n.o.^T}, \mathbf{x}_1\left(r\right)^{n.o.^T}\right]^T \qquad (5.17)$$

$$\Delta\mathbf{x}^e\left(r\right) = \left[\Delta\mathbf{x}_{-1}\left(r\right)^T, \Delta\mathbf{x}_0\left(r\right)^T, \Delta\mathbf{x}_1\left(r\right)^T\right]^T \qquad (5.18)$$

and where the superscripts '*ort.*' and '*n.o.*', which stand for 'orthogonal' and 'non orthogonal', respectively, denote the vectors that meet the orthogonality condition with respect to the spreading code $\mathbf{c}$ and those that do not. The error signal in (5.14), briefly denoted as

$$\mathbf{e}^e\left(r\right) = \gamma b(r)\mathbf{y}^{e*}\left(r\right) = \left[\mathbf{e}_{-1}\left(r\right)^T, \mathbf{e}_0\left(r\right)^T, \mathbf{e}_1\left(r\right)^T\right]^T, \qquad (5.19)$$

is built up as sketched in Figure 5-11. The vector $\mathbf{x}^e$ is made orthogonal to the code $\mathbf{c}$, according to (5.15), where the correction term $\Delta\mathbf{x}_w$ is defined as

$$\Delta\mathbf{x}_w\left(r\right) = \left(\frac{\mathbf{x}_w\left(r\right)^{n.o.^T}\mathbf{c}}{L}\right)\mathbf{c} \qquad (5.20)$$

and is computed by the circuit block detailed in the following.

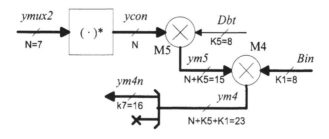

Figure 5-11. EC-BAID error signal generation.

Implementation of (5.14) and (5.15) requires a few memory elements for storing the $\mathbf{x}^e$ vector. As shown in Figure 5-12, the architectural choice was to store the 384×46 bits of the vector $\mathbf{x}^{e\,n.o.}$ in an SRAM, before orthogonalization (5.15). Internal arrangement of the SRAM is shown in Figure 5-13. The correction term $\Delta\mathbf{x}_w$, which is subtracted from $\mathbf{x}_w^{n.o.}$ to yield the vector $\mathbf{x}_w^{ort.}$ (orthogonal to the code $\mathbf{c}$), is generated by the block labeled '*ortog.*' in Figure 5-12, which creates the final $\mathbf{x}^{e\,ort.}$ together with the subtractor S4. Vector upgrading is implemented according to (5.14) by

Vector upgrading is implemented according to (5.14) by means of the sub-tractor S3, and the EC-BAID adaptation step is coded as a power of 2 in order to allow the use of a shift in the place of a multiplier.

As reported in Chapter 3, a significant performance improvement with respect to the original algorithm was obtained after insertion of a 'leakage' correction, in order to cope best with timing bias effects. According to such a correction, (5.12) becomes

$$\mathbf{x}_w(r+1) = \mathbf{x}_w(r)(1-\gamma F_l) -$$
$$-\gamma b(r-1)\cdot\left[\mathbf{y}_w^*(r-1) - \frac{\mathbf{y}_w^*(r-1)^T \mathbf{c}}{L}\mathbf{c}\right], \tag{5.21}$$

where F_l is the leakage factor. In the architecture selected for the ASIC, the modified version of (5.14) featuring the leakage correction factor, is

$$\mathbf{x}^e(r+1)^{n.o.} = \mathbf{x}^e(r)^{ort.}(1-\gamma F_l) - \gamma b(r-1)\mathbf{y}^{e^*}(r-1), \tag{5.22}$$

which can be re-written as

$$\mathbf{x}^e(r+1)^{n.o.} = \mathbf{x}^e(r)^{ort.} - \gamma\left[F_l\mathbf{x}^e(r)^{ort.} + b(r-1)\mathbf{y}^{e^*}(r-1)\right], \tag{5.23}$$

leading to the implementation in Figure 5-12.

We notice that all operations described in this Section are performed on extended $3L$-element vectors by re-using the same arithmetical resources three times per chip period.

2.1.5   Input and Code RAM

The arithmetic blocks described in sub-Sections 2.1.2 and 2.1.5 process the input chip values of current rth symbol and of previous $(r - 1)$th, $(r - 2)$th and $(r - 3)$th symbols. This is accomplished by storing the most recent $3L$ input chips in the SRAM of Figure 5-14, which also stores the code sequence bits loaded during the initialization phase. Appropriate generation of RAM addresses together with a consistent control of the multiplexers in Figures 5-12 and 5-14 ensures the correct timing for the arithmetic resources that have to be used three times for each chip period, as detailed in the timing diagram of Figure 5-15. Similar waveforms are reported in Figure 5-16 for the main symbol rate signals involved in the EC-BAID implementation.

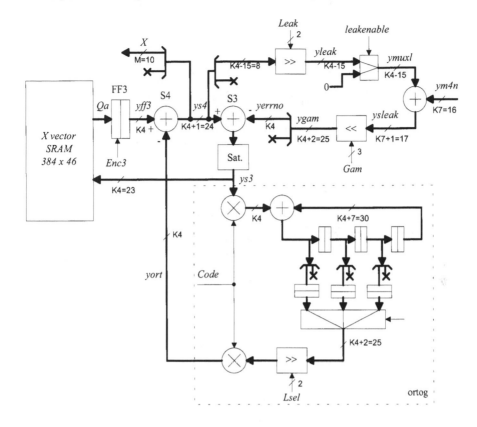

Figure 5-12. $\mathbf{x}^e$ vector upgrade.

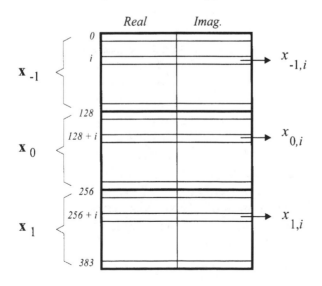

Figure 5-13. $\mathbf{x}^e$ vector SRAM arrangement.

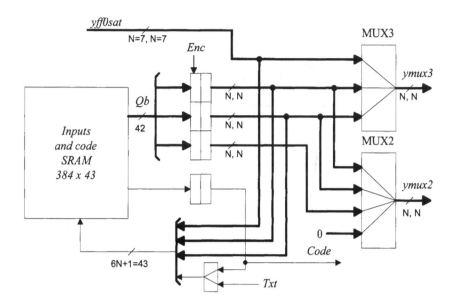

Figure 5-14. Input RAM and multiplexers.

As shown in Figure 5-15, the chip period is subdivided into eight sub-periods according to the value of the internal state *CS*. Most arithmetical operations occur three times per chip, each within a $T_c/4$ period selected according to the *msel* control value. The last chip quarter (S6 and S7 states) is an idle time which is occasionaly skipped if the CCTU needs to anticipate the symbol starting reference. The input RAM (RAM_Y) is accessed two times per chip (to perform one read and one write operation), while the RAM of the vecor $\mathbf{x}^e$ (RAM_X) is accessed six times per chip, in order to read and write three elements values. The RAM enable (active low) control signals are *csna* for RAM_X and *csnb* for RAM_Y, while *rwn* sets the write (0) or read (1) cycle for both memories. The least significant portion (*addr[6:0]*) of RAM_X addresses (*addr[8:0]*) is also used for RAM_Y.

The timing diagram in Figure 5-15 is related to the calculation of the output $b(r)$ for the rth symbol, that takes place while the ASIC is receiving the chip rate samples relevant to the $(r + 1)$th symbol. In this example, the chip under process is the ith one. Figure 5-17 depicts the contents of RAM_Y: words with indices from 0 to $i-1$ have already been updated, whilst rows with indices between i and L are holding the old values.

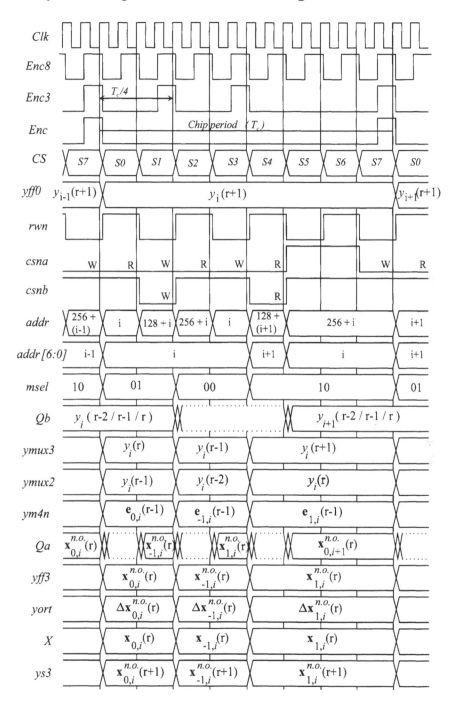

Figure 5-15. Main chip rate signals' timing diagram.

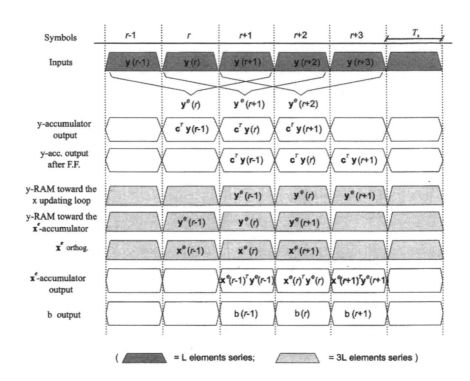

Figure 5-16. Main symbol rate signals timing diagram.

2.1.6   Carrier Phase Recovery Unit

The CPRU block performs carrier phase recovery on the symbol rate signal strobes at the detector output. It is based on a second order Costas loop according to the equations detailed in Chapter 3, and includes a lock detector also described in the same Chapter. In particular, the CPRU performs two complex rotations per symbol time (for both the EC-BAID and the standard correlator outputs) by using the same multiplier eight times per symbol period. As depicted in Figure 5-18, the CPRU can be bypassed (to use external carrier phase recovery) by setting to 0 the *costasonoff* control parameter, which selects between the *ycr* and *ybaid* symbol rate signals and their counter-rotated versions. The hardware multiplexing introduced for the phase rotator block needs some registers in order to separate and re-align the EC-BAID and standard correlator outputs (*CRout* and *Bout*).

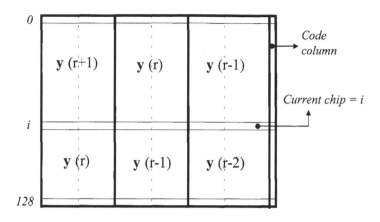

Figure 5-17. Input RAM organization

2.1.7   Output Management

Final stages in the outputs control path are controlled by *out_sel* and *swap_sel* parameters as detailed in Section 1.1 (see Tables 5-2 and 5-3) and in Figure 5-19. The auxiliary signal *modulox* is computed by a block that, during every symbol period, accumulates the modulus of the vector $\mathbf{x}^e$ to build the norm of the adaptive vector. Specifically, a simplified implementation like the one described for the AGC is utilized in order to have low complexity estimation of the vector norm X [Fan02].

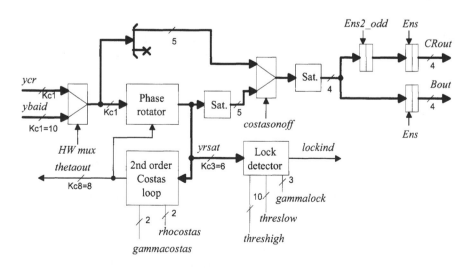

Figure 5-18. CPRU block diagram.

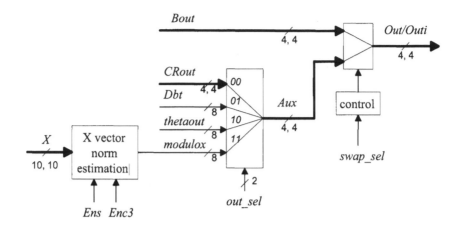

Figure 5-19. Output control.

2.1.8 Control Blocks

Three ASIC control blocks, implemented as finite state machines, are responsible for the correct circuit timing.

The main synchronization block regulates the initialization phases, as illustrated in Figure 5-4. In the first phase (state = *RESET*), which occurs when the Resn input is low, all the registers (both in the arithmetical paths and in the control state machines) are forced to 0, so no enable strobes are produced and the SRAMs are not accessed. When Resn goes high the parameter loading phase (state = *REC*) is started by activating a second control block, which works with the Req, Txt and Rack I/O signals as illustrated in Figure 5-5. The duration of this phase depends on the external response time (Txt and Rack coming from the DSP in the MUSIC receiver breadboard), and concerns the transmission of $57 + L$ bits. Code sequence bits are stored in the last column of the input RAM (RAM_Y) during this phase, as sketched in Figure 5-14. In order to start the algorithm from a known state, a third initialization phase takes place (state = *ERASE*) during which the vector $\mathbf{x}^e$ RAM and input RAM words are loaded with 0 values (all the columns but the code sequence one). Finally, normal operations phase is started (state = *OP*), and circuit control is passed to the *enable generator* block. The main synchronization sequence is thus completed, and it will restart from the *RESET* state when another low pulse arrives on the Resn input.

The enable generator waits for a Sym_in pulse and produces all the enable strobes, RAM addresses, and hardware multiplexing control signals for the current symbol period; then, it waits for the next Sym_in pulse. If no symbol references are received the circuit is idle, and power consumption is reduced. On the other hand, if Sym_in anticipates, signal generation is immediately restarted skipping the last portion of the symbol period. Together with the presence of an idle time at the end of each chip period (states S6 and S7 in Figure 5-15) this behavior allows for correct symbol synchronization even when the CCTU needs to re-align the symbol starting time. A timing diagram with the main enable and control signals generated by this block is shown in Figure 5-20.

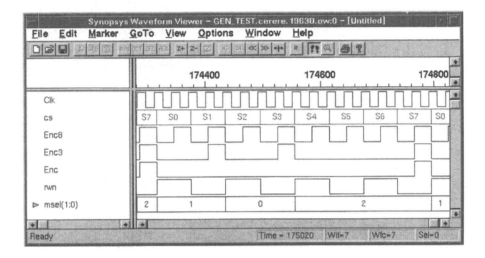

Figure 5-20. Enable generator main signals.

3. ASIC IMPLEMENTATION

Once the FPGA validation steps described in Chapter 4 were accomplished, the RTL architecture of the EC-BAID presented above was used as the starting point for the ASIC implementation design flow (which is the topic of this Section), starting from VHDL down to final layout. As a matter of fact, the design flow is exactly the same as the one sketched in Figure 4-2, apart from some necessary modifications in the Back End phase caused by the different target technology. After a brief overview of the target technology, this Section describes all the Front End and Back End phases performed in the ASIC design flow.

3.1 Technology Overview

The ASIC implementation was targeted to the 0.18 μm CMOS technology HCMOS8D, provided by STMicroelectronics. Some of its main features are reported in this Section, together with additional information about the selected package.

3.1.1   The HCMOS8D Technology and Its Relevant Design Libraries

The HCMOS8D technology, provided by STMicroelectronics, is based on a CMOS process whose main features as the following [hcmos]:

- 0.18 μm minimum gate length, obtained through deep Ultra Violet (UV) lithography;
- 6 levels of metal, which ease the routability of very high density circuits;
- Fully stackable vias and contacts;
- Thin gate oxide (35 Angstrom);
- Shallow Trench Isolation between active regions, improving transistor density and planarity;
- Salicided active areas and gates (to yield lower resistance).

The version of the standard cell library CORELIB (from the CB65000 family) is the one optimized for low leakage and 1.8 V power supply, contains over 750 cells, ensures an average gate density of 85 kgates/μm^2 and a 60 ps delay for a typical NAND2 gate. PAD libraries IOLIB_50 and IO-LIB_80 (again from the CB65000 family) offer a wide range of I/O pad types, with different versions for pad widths of 50 μm or 80 μm and for core power supplies of 1.3 V or 1.8 V. The selected IOLIB_80 version contains over 430 cells, each with ESD and latch up protections, and is intended to interface with 1.8 V core logic. I/O pads can interface with 1.8 V external signals, and a complete set of 3.3 V capable inputs/outputs is also included, together with 5 V tolerant inputs. Several options are also available for input, output or bidirectional I/Os: compensated active slew-rate, pull up and pull down capability, split ground, integrated test pins, Schmitt trigger inputs, tri-state outputs, different drive strengths.

Several types of RAM/ROM cuts can be obtained by means of automatic memory generators, as well as Built In Self Test (BIST) blocks.

3.1.2   Package Selection

As mentioned in Section 1.1, the small size of the EC-BAID circuit suggested to reduce the pad number down to 44 pin, in order to limit the amount of unused silicon in the pad limited ASIC layout. Given this pin number and given an estimated die area lower than 2×2 mm^2, the smallest available package, the TQFP4410x10, was selected. Its external area is 10×10 mm^2, while the internal cavity has a width of 6 mm^2, the smallest available for this family.

3.2 Front End Design Flow

The bit true architecture described in the previous Chapter was validated and optimized by means of several simulation runs, resulting in a final choice for all the bus widths. Such an RTL circuit description was the starting point for the subsequent Front End design flow phases, that are sketched in Figure 5-21 and are described in this Section.

3.2.1 VHDL Description

The first step in the Front End flow was the description of the EC-BAID architecture in VHDL. The hierarchical structure was described in a mixed structural and behavioral style. More in detail, for the circuit data paths, the VHDL high level arithmetical operators were mixed with explicit register instantiations, while the finite state machines in the control blocks were implemented as VHDL processes. The whole design is fully synchronous to simplify all the project phases. Design for testability issues were taken into account by including Built In Self Test (BIST) blocks for the RAMs, as well as automatically generated scan chains for the rest of the circuit.

The memory blocks size were defined considering the diverse available automatic generators (SPS4, SPS2HD, SPS6, SP8D, DPR2, DPR8D, etc.) [hcmos] and the different amount of control logic. As a final choice, an implementation of two separate single-port synchronous RAMs was decided, allocating the SPS4 generator for the 128×43 cut and the SPS2HD for the 384×46 cut. SPS4 is targeted for small sizes and low power, while SPS2HD produces high speed and high density RAMs. Since the timing constraints did not reveal critical in this design, the best choice for this ASIC would have been the SPS4 for both memories. However, the 384×46 size exceeded the maximum word number allowed for SPS4, and therefore the SPS2HD generator was selected for this cut.

Two BIST blocks were embedded in the ASIC, one for each RAM, in order to improve the overall testability. Their function is to provide RAM fault detection by keeping memory blocks separate from the scan paths. This prevents creation of poorly observable and controllable sections of the circuit. BIST operations are controlled by the two input pins Bact and Tm, according to Table 5-7.

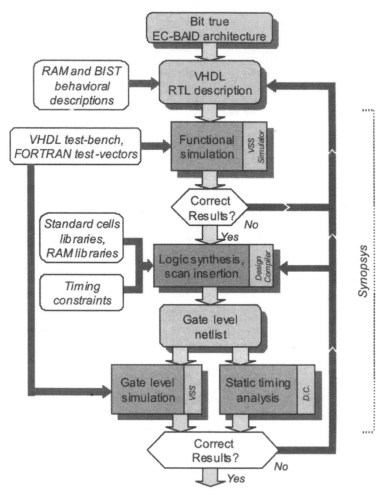

Figure 5-21. Front End design flow.

Table 5-7. BIST modes and control pins.

Bact	Tm	Operation
0	0	*Transparent Mode* : System logic drives RAM
1	0	*Self Test Mode* : BIST controls RAM
0	1	*Test Mode* : RAM bypass, scan path test
1	1	*Scan Collar Test Mode* : not implemented

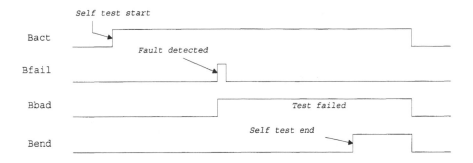

Figure 5-22. BIST signals sequence.

During the self-test mode, the BIST block executes the automatic memory test, while in normal test mode the RAM is bypassed and the BIST itself is tested with the rest of the logic in the scan path. As shown in Figure 5-22, the self-test is started when the Bact input signal goes high, and terminates when Bend is set. Whenever an error is detected a pulse appears on Bfail, so that the Bbad signal is set high and held to that value until the test is over. The overall duration of the testing algorithm (named Marinescu 17N) in clock periods is

$$N_{test} = 17 \cdot N_{words} \cdot \left[\log_2(N_{data}) + 1 \right]. \tag{5.24}$$

The test detects several types of faults (stuck at 0/1 cells, disconnected cells, cell coupling, decoder malfunctioning, etc.). Each one of the two BIST blocks has three test output signals (Bfail, Bbad and Bend). In order to keep the number of ASIC pad low, these six test output signals are assigned to six of the standard output pads by means of a multiplexer controlled by the Bact signal, as detailed in Table 5-8.

Table 5-8. BIST outputs assignments.

Pad number	Standard output (Bact = 0)	BIST output (Bact = 1)
20	Outr_2	Bend(X)
21	Outr_1	Bbad(X)
22	Outr_0	Bfail(X)
25	Outi_2	Bend(Y)
26	Outi_1	Bbad(Y)
27	Outi_0	Bfail(Y)

The VHDL description of the whole circuit, enclosed in a top level block, was functionally tested via the VSS™ simulator by Synopsys with no timing information. An appropriate VHDL test bench was developed in order to make a comparison between the simulated circuit outputs and previously produced FORTRAN bit true test vectors. As depicted in Figure 5-23, several Boolean flags monitor matching with the test vectors. Also, the test bench simulates possible timing realignments imposed by the CCTU. The example in Figure 5-24 shows a symbol reference (Sym_in) shifted by $T_c/4$ (in advance). The internal enable signals (Enc3 and Enc) are immediately re-aligned and the EC-BAID outputs continue to match the test vectors.

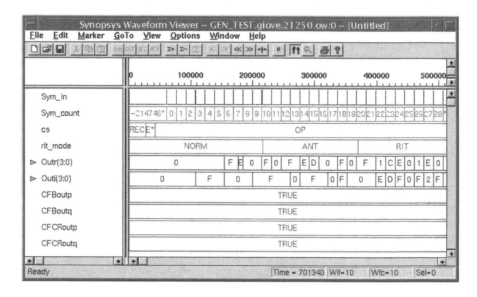

Figure 5-23. RTL simulation example.

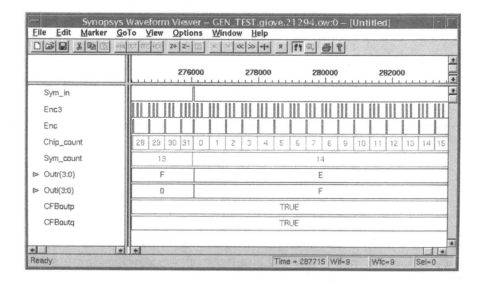

Figure 5-24. RTL simulation detail.

3.2.2   Circuit Synthesis

After RTL simulation and validation, logic synthesis of the core circuit was carried out via the Design Compiler™ tool by Synopsys [Bha02]. Since the timing was not very critical, synthesis constraints were set allowing priority to area and power saving issues. The clock frequency was set to 40 MHz in order to have a safe margin for the maximum required operating frequency of 32.768 MHz (which allows a maximum chip rate of 4.096 Mchip/s). Many multi-cycle path declarations were inserted to correctly address the synthesis of arithmetical blocks operating at different rates (from data symbol rate R_s up to circuit internal clock rate $8 \cdot R_c$), while the clock was marked as ideal in order to synthesize a balanced clock tree in the Back End phases.

The main synthesis results, as summarized in Table 5-9, show that the total area of cells and memories is small, leading to the envisaged pad limited layout with non-critical placement and routing operations.

Table 5-9. Synthesis results.

Total area (RAM + standard cells)	0.676 mm^2
RAM area	0.281 mm^2
Standard cells number (without clock tree)	13622
Fault coverage	99.24 %
Maximum frequency	40 MHz

The most critical path is a two clock period multi-cycle path starting from the FF1 register of Figure 5-8 and ending to a register of the *ortog* block in Figure 5-12, which meets the timing requirements with a 0.12 ns positive slack. After the synthesis phase, which included a hold violation fixing step, the circuit revealed compliant with all of the timing constraints, and a gate level output netlist was produced for the next operations.

A gate level pre-layout simulation was then performed on this netlist to validate the circuit behavior. The synthesized netlist was input to the VSS™ simulator, together with the timing back-annotations (in Standard Delay Format — SDF), using the previously developed test bench. A simulation screenshot is reported in Figure 5-25, where a multi-cycle arithmetical path with a 28 ns delay is shown as an example. In particular, this picture shows the non ideal delays of the internally generated enable signals (Enc8, Enc3, Enc) and the propagation delay (28 ns) of an internal signal (inf-fincl[59:0]) at the end of a long combinatorial path. The Boolean flags confirm the correct behavior of the circuit.

After the first optimization step, the scan chain was created by connecting all of the 1017 scan flip flops in a unique scan path. The test vectors that were automatically generated for this chain ensure a fault coverage over 99%, as shown in Table 5-9.

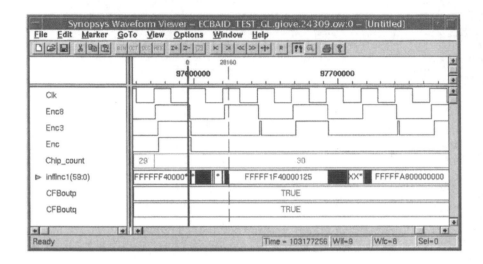

Figure 5-25. Gate level pre-layout simulation detail.

3.3   Back End Design Flow

The Back End flow is sketched in Figure 5-26. Once the PAD checks were carried out with the ICpack tool, all the *place and route* phases were accomplished via the CAD tool Blast Fusion™ by Magma. The post-layout verification steps were instead performed with industry standard tools by Synopsys, Cadence and Mentor [Smi97].

3.3.1   PAD Selection

Recalling the I/O signals and package description of Sections 1.1 and 3.1.2, we observe that the ASIC has 44 pins, with 35 I/Os and 9 power supply pads. The first choice to be made was the selection of the pad library between IOLIB_80 and IOLIB_50 (which contain pad cells of size 80×220 μm^2 and 50×380 μm^2, respectively). Although IOLIB_50 is the library intended for pad limited circuits, the geometrical constraints for this ASIC suggested the use of the 80 μm pad version. As illustrated in Figure 5-27, the IOLIB_80 pads lead to the minimum side length of 1320 μm.

I/O pads selection was made by choosing simple pads with 3.3 V capability in order to ease interfacing with the FPGA of the MUSIC receiver breadboard. All the input signals use the TLCHT_TC pad cell (the basic 3.3 V capable input), with the exception of Resn, which utilizes a SCHMITT_TC pad cell (with a Schmitt trigger included) in order to speed up possible up possible asynchronous transitions. The pad cell used by all output pins is the B2TR_TC, a 3.3V output pad with slew-rate control and a maximum DC current of 2 mA, suited for loads up to 50 pF.

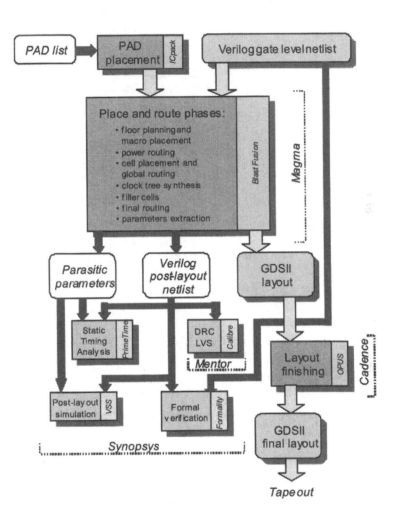

Figure 5-26. Back End design flow.

Identification of the correct number of power supply pads calls for power consumption estimation. This was accomplished following proper guidelines provided by the silicon foundry. A first instance, rough power estimate was quickly calculated by Synopsys Design Compiler, which can combine the registers switching activity monitored during an RTL simulation with statistically estimated activities for the remaining combinatorial cells. This method resulted in an estimate of about 12 mW for the core power consumption, at a clock speed of 32.768 MHz, and with a chip rate of 4.096 Mchip/s.

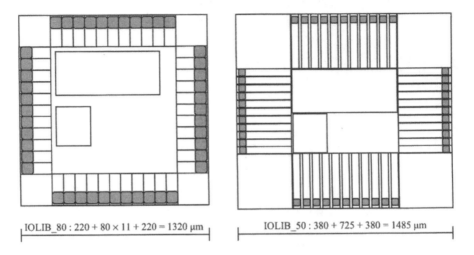

IOLIB_80 : 220 + 80 × 11 + 220 = 1320 μm   IOLIB_50 : 380 + 725 + 380 = 1485 μm

Figure 5-27. Die area with different pad libraries.

According to the above mentioned guidelines, 2 VDD3IOCO pads were inserted in order to provide the 3.3 V power supply to all I/O pads, whilst 2 VDDIOCO pads were included to provide the 1.8 V power supply for the core and the internal I/O cells buffers. Moreover, 5 VSSIOCO ground pads were put in the remaining places. All I/O and supply pads include Electro-Static Discharge (ESD) protections, ruling out the need for specific cells.

Pad cells were added to the netlist after the logic synthesis, while their placement was performed as the first Back End step by means of the ICpack tool. This software placed the pad cells taking the desired order into account (as in Figure 5-1), and checking all the packaging rules. Its output was a Physical Design Exchange Format (PDEF) file, which is a proprietary file format used by Synopsys to describe placement information and clustering of logic cells. Supplementary spaces were added between the most peripheral pads and the corner cells in order to avoid bonding rules violations. This resulted in a final die area of 1528×1528 μm² with the IOLIB_80 pads. Starting from Figure 5-27, and considering this added length and the amount of space necessary for RAM buses routing, the 80 μm pad library still revealed the correct choice.

In order to avoid the simultaneous switching of all the output pads, which could impair power supply levels, additional delay cells were inserted between final registers and Outr/Outi output pads to provide a set of different delays (however negligible with respect to the output signals symbol rate).

3.3.2   Place and Route Flow

The whole set of Back End phases, from the synthesized gate level netlist to the GDSII, were performed by means of Blast Fusion™ by Magma. This tools was selected because it addresses circuit timing closure in a different, more efficient way with respect to competing products available on the market (for example, the widespread Silicon Ensemble™ by Cadence). Since wire delays are becoming the predominant delay factor, a design flow that executes placements for optimized area and then performs the routing according to the timing constraints may require several iterations and re-optimization phases. On the contrary, the design flow proposed by Magma Blast Fusion™ addresses the timing closure problem from the very first phases, exploiting the proprietary *FixedTiming* methodology together with the *SuperCell* approach. Magma's FixedTiming methodology combines logical and physical design to ensure better performance by eliminating iterations between synthesis and 'place and route' phases. With FixedTiming, Blast Fusion™ determines the optimal timing of the design prior to detailed routing. The system then dynamically controls the size, placement and wire interconnects of each cell to preserve the established optimal timing. This 'correct by construction' approach eliminates the need to re-synthesize to improve on bad timing performance.

To achieve optimal timing, each logic cell must have the proper drive strength for the relevant load. Because interconnect delay cannot be determined or accurately estimated during synthesis, Magma continually varies cell sizing during place and route to maintain constant timing. Rather than using pre-sized cells from the target library, Magma replaces each logic function with automatically abstracted SuperCell models (functional placeholder cells with variable sizes and fixed delay, as sketched in Figure 5-28). Initial placement and routing of the SuperCells allows Magma to determine the final optimal timing for all paths in the design. The layout is then completed by continuously adjusting the size of each SuperCell so that the delay stays constant. Finally, the SuperCells are replaced with actual library cells that have the proper drive strength. As sketched in Figure 5-29, all the place and route tasks take place in the same tool, allowing the use a single unified data model which is very useful for the management of large size chips.

The Verilog synthesized gate level netlist, the pad placement PDEF file, the timing constraints set, as well as every needed library database were then the inputs to the Blast Fusion tool. The first step accomplished within the Magma tool was the definition of an initial floorplan with RAM blocks placement, followed by the creation of a power routing grid in metal 5 and metal 6. Then the cell placement, the clock–tree synthesis, and the final routing were performed with the previously described methodologies, obtaining the whole ASIC layout in GDSII format. A final parasitic parameters extraction was performed to obtain a Standard Parasitic Format (SPF) file for additional post-layout timing analysis.

The resulting output from Blast Fusion flow was the GDSII layout, the SPF parasitic parameters, a final Verilog post-layout netlist, and the related timing exception set in Synopsys Design Constraint (SDC) format.

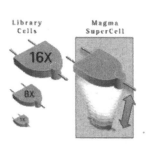 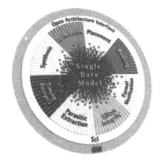

3.3.3   Post-Layout Checks

After the different phases described above, several post-layout checks were carried out by means of different tools. A static timing analysis was carried out using Synopsys PrimeTime™, which read back the final netlist with the extracted parasitic parameters in order to check all circuit timing requirements. A formal verification was then made with Formality™ by Synopsys to ensure the logical equivalence between the starting gate level netlist and the final post-layout netlist.

Layout checks were performed with Calibre™ by Mentor, consisting in a Design Rule Check (DRC) step to control the absence of design rule violations, followed by a Layout Versus Schematic (LVS) step to check the correspondence between the final gate level netlist and the actual layout.

All these final checks were correctly passed, together with a very last Synopsys VSS™ gate level simulation.

3.3.4   Layout Finishing

Before tape out a final step was performed with Cadence OPUS™ to insert all the additional elements needed by the foundry in the GDSII, like alignment patterns, mask identification numbers, logos and external scribe lines. A view of this final layout is shown in Figure 5-30, whilst the packaged component plugged on the board to be connected to the Proteo I board is shown in Figure 5-31.

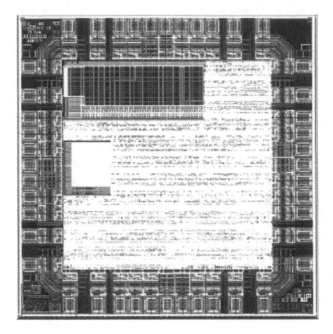

Figure 5-30. Final EC-BAID ASIC layout.

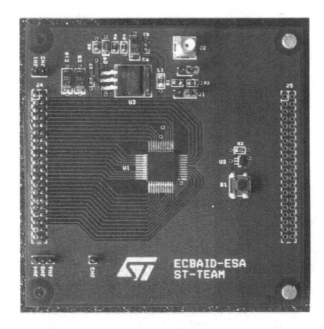

Figure 5-31. EC-BAID ASIC mounted on the board to be connected to the PROTEO board.

3.3.5   Design Summary

Some of the main ASIC features before packaging are listed below.

- **Area**: the final ASIC size is 1528 μm×1528 μm = 2.33 mm^2.

- **Speed**: the worst case timing analysis reports a maximum allowed frequency of 40 MHz, which implies a maximum chip rate of 5 Mchip/s. The range of chip rates envisaged by the MUSIC project is thus fully covered.

- **Power**: a final power estimation resulted in a total power consumption of 12.5 mW at the clock frequency of 32.768 MHz, with a chip rate of 4.096 Mchip/s, which is twice the maximum chip rate specified for the MUSIC project.

- **I/O timing**: the setup/hold timing requirements for all the input signals with respect to the clock rising edge arrival time at the Clk pin, as extracted by the PrimeTime analyzer, are reported in Table 5-10. Output delays in the case of 20 pF external loads are listed in Table 5-11.

Table 5-10. Input timing requirements.

Input pin	Setup time (ns)	Hold time (ns)
Sym_in	0.0	0.82
Resn	0.0	0.48
Enc8	0.0	0.82
Yr	0.61	0.39
Yi	0.59	0.50
Txt	3.09	0.71
Rack	0.0	0.73
Bact	6.30	0.38
Tm	5.33	0.12
Test_si	0.10	0.64
Test_se	3.42	0.64

Table 5-11. Output delays with 20 pF loads.

Output signal	Max. delay time (ns)
Req	10.16
Sym_out	11.81
Lock	24.17
Outr_3	13.20
Outr_2	13.68
Outr_1	14.94
Outr_0	16.07
Outi_3	16.35
Outi_2	18.65
Outi_1	20.34
Outi_0	22.07

Chapter 6

TESTING AND VERIFICATION OF THE *MUSIC* CDMA RECEIVER

We describe in this chapter the real time testbed facility that was set up to validate the MUSIC receiver, from the features of signal, interference and noise generation down to the hardware architecture and the ultimate receiver performance. The ultimate purpose of the testbed was actually two-fold: on the one hand it helped debugging the MUSIC receiver (thus getting rid of any possible implementation bug); and tuning the diverse loop parameters. On the other, it allowed us to carry out the Bit Error Rate (BER) performance characterization in a synthetic environment that closely mimics the features of a typical satellite communication downlink.

1. REAL TIME TESTBED DESIGN

1.1 Overall Testbed Architecture

Repetita iuvant (repeating helps) used to say our Roman ancestors, so we state once more that the ultimate goal of the MUSIC experiment was to validate, through a proof of concept breadboard, a single-ASIC implementation of the EC-BAID detector, as well as to demonstrate the suitability of the whole receiver to integration into a hand held user terminal. Picture 6-1 offers a view of the MUSIC testbed built up at the project facility center [Fan01]: the master PC and several pieces of instrumentation, including the digital boards accommodating the receiver, can be easily identified. The actual architecture of the testbed is sketched in Figure 6-2, and its main features are listed hereafter:

1. Flexible and programmable generation of the useful plus interfering CDMA signal;
2. Injection of Gaussian noise with programmable level;
3. Analog IF interface between the signal generator and the MUSIC receiver test board;
4. Interface of the MUSIC receiver to subsequent baseband processing (e.g., BER measurement, optional error correcting decoding, etc.);
5. Monitoring capabilities;

Signal plus Multiple Access Interference (MAI) generation is performed via a computer controlled arbitrary waveform generator, followed by frequency upconversion to the standard analog intermediate frequency 70 MHz, and by injection of Additive White Gaussian Noise (AWGN) performed with the aid of a precision noise generator. A master PC controls the testbed via IEEE488 bus by means of a dedicated program specially developed in LabVIEW. On one hand this improves configuration controllability and system flexibility; on the other performance results in terms of BER (Bit Error Rate), internal signals spectra monitoring, sync parameters evolution and so on are easily attained.

Figure 6-1. A corner of the MUSIC lab.

The MUSIC receiver consists of two sections, namely an IF analogue front end, and a digital platform hosting the digital signal demodulator. The latter is composed of two separate boards: a digital breadboard named PROTEO, which is intended to accommodate the digital receiver front end, as well as the slower rate ancillary functions of synchronization and house-keeping, and a plug in mini board supporting the single ASIC implementation of the EC-BAID detector [MUS01].

The analog IF front end performs IF channel filtering via an appropriate SAW filter, and signal amplitude automatic control to regulate the total received power as well as a suitable level for the subsequent Analog to Digital Converter (ADC) mounted on the digital breadboard.

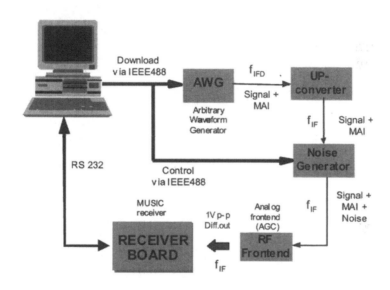

Figure 6-2. MUSIC testbed architecture.

The digital section of the receiver is shown in Figure 6-3, which displays the PROTEO breadboard implementing the MUSIC receiver, along with the plug in board hosting the ASIC of the EC-BAID detector.

As mentioned above, the MUSIC receiver building blocks that are ancillary to the EC-BAID detector were implemented in the PROTEO breadboard, a programmable platform specifically designed by STMicroelectronics [MUS01] and whose functional block diagram is sketched in Figure 6-4.

The digital computational capability of the PROTEO breadboard mainly relies on two Complex Programmable Logic Devices (CPLD) equipped with 100 kgates each, and provided by Altera[TM]. These devices contain programmable SRAM memory that is re-configurable when in the circuit, either via an external connector (Bit-Bluster) or by internal Flash memory. Each device also contains 624 logic units, or logic array blocks (LAB) with 8 basic logic elements each (LE), and 24 kbit RAM memory arranged in 12 embedded array blocks (EAB). The LABs are used to implement combinatory functions such as adders, multiplexers or sequential elements, while EABs are mainly used either for storing purpose, as for RAM and ROM, or for implementation of complex functions.

Figure 6-3. Picture of the PROTEO DSP board with the EC-BAID
ASIC mini-board (upper left).

To increase system controllability and flexibility, the breadboard is also provided with a high performance ST18952 DSP processor, operating in 16 bit fixed point arithmetic, with a worst case speed of 66 Mips/15 ns; the ST18952 is equipped with 32K words program memory and 16.5K words data memory.

Thanks to the proper configuration of a set of 12 bit high speed tri-state buffers, the breadboard can be fed either via a digital input connector, or via two ADC converters (ADS807), both interfaced to the first CPLD.

The master clock of the board is generated by a VCXO oscillator that acts as the master frequency reference for a clock buffer/generator component with programmable skew outputs (CY7B991). The latter generates five separated clocks at 16.384 MHz that are user-controllable skewed (± 6 time-units) by a hard wired, pull up or pull down, set of resistors.

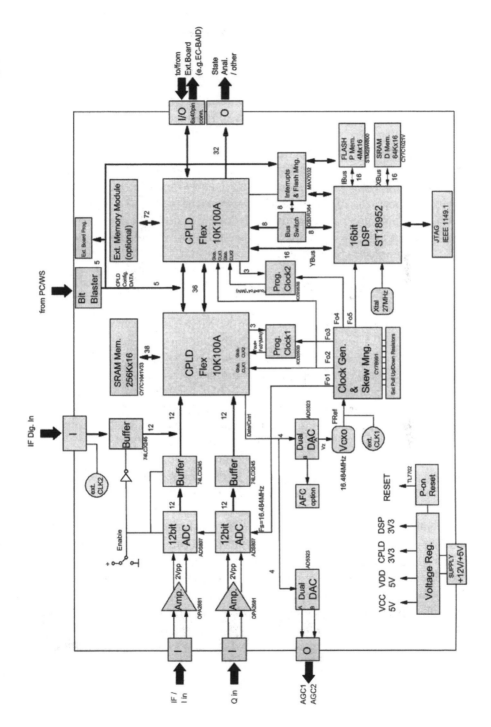

Figure 6-4. PROTEO breadboard functional block diagram.

Moreover, additional programmable clock generators (ICD2053B, digital PLLs) allow the generation of any clock frequency in the range 391 KHz to 90 MHz 'on the fly'.

A large amount of memory, for general purpose processing, is available to both CPLDs. A 256 kbit SRAM chip is connected to the first one, whilst a SIMM-like connector for a SRAM 1MB plug in module is connected to the second.

A comb of 40 pin headers encircles both CPLDs, allowing digital signals monitoring, as well as external I/Os connection for additional 'plug in' extension boards (for example, the smaller board where the EC-BAID ASIC device is mounted). Two more 40 pin headers, connected to the second CPLD, are also compatible with the HP5600 State/Logic analyzer probes. Finally a set of chips provides regulated levels of voltage in the range 3.3-5 V to supply the breadboard.

1.2   CDMA Signal Generation

The CDMA signal for the testing of the MUSIC receiver is generated as follows [MUS01]. First, a FORTRAN computer simulation is run off-line in order to provide a properly sampled version (with floating point amplitude resolution) of the desired waveform spanning a given number of symbol intervals. The sampling frequency of the simulated signal is $f_s = 16.384$ MHz, which keeps some degree of symmetry between transmit and receive clock speeds. The available bit and chip rates of the CDMA signal are shown in Table 6-1.

The parameters of the CDMA signal are passed to the FORTRAN program by means of a friendly Graphical User Interface (GUI), suitably developed using the National Instruments' LabView environment. Such a solution allows for a quick and easy re-configuration of the test signal parameters, thus yielding a maximum of flexibility. The GUI outputs a file containing all the parameters of the CDMA signal, and such file is used as input by the FORTRAN simulator. The simulation program, in addition to generating a pseudo-random bit sequence for the useful channel bit stream, also performs frame formatting. In particular it adds a pattern of 24 QPSK symbols (48 bits), the so called Unique Word (UW), at the beginning of the simulated waveform for frame synchronization purposes at the receiver side.

The FORTRAN program outputs two files: a binary file containing the signal samples (represented on 16 bits, fixed point) to be handled by the Arbitrary Waveform Generator (AWG), and an ASCII text file containing the stream of the transmitted information symbols, to be used jointly with the data estimates provided by the receiver for BER measurement. The waveform obtained by computer simulation is a CDMA signal compliant with al

MUSIC specifications and modulated onto a first Intermediate Frequency (IF). Since the signal is in digital form such a frequency is referred to as Digital IF (IFD), and is set to $f_{IFD} = 4.464$ MHz (see Figure 6-5).

Table 6-1. Values of R_c, (kchip/s) R_b (kbit/s) and L for the MUSIC signal.

R_b	n	$L = 32$	n	$L = 64$	n	$L = 128$
2	4	$R_c = 128$	2	$R_c = 128$	1	$R_c = 128$
	8	$R_c = 256$	4	$R_c = 256$	2	$R_c = 256$
	16	$R_c = 512$	8	$R_c = 512$	4	$R_c = 512$
	32	$R_c = 1024$	16	$R_c = 1024$	8	$R_c = 1024$
	64	$R_c = 2048$	32	$R_c = 2048$	16	$R_c = 2048$
4	2	$R_c = 128$	1	$R_c = 128$		$R_c = $ ---
	4	$R_c = 256$	2	$R_c = 256$	1	$R_c = 256$
	8	$R_c = 512$	4	$R_c = 512$	2	$R_c = 512$
	16	$R_c = 1024$	8	$R_c = 1024$	4	$R_c = 1024$
	32	$R_c = 2048$	16	$R_c = 2048$	8	$R_c = 2048$
8	1	$R_c = 128$	--	---	--	---
	2	$R_c = 256$	1	$R_c = 256$	--	---
	4	$R_c = 512$	2	$R_c = 512$	1	$R_c = 512$
	8	$R_c = 1024$	4	$R_c = 1024$	2	$R_c = 1024$
	16	$R_c = 2048$	8	$R_c = 2048$	4	$R_c = 2048$
16	1	$R_c = 256$	--	---	--	---
	2	$R_c = 512$	1	$R_c = 512$	--	---
	4	$R_c = 1024$	2	$R_c = 1024$	1	$R_c = 1024$
	8	$R_c = 2048$	4	$R_c = 2048$	2	$R_c = 2048$
32	1	$R_c = 512$	--	---	--	---
	2	$R_c = 1024$	1	$R_c = 1024$	--	---
	4	$R_c = 2048$	2	$R_c = 2048$	1	$R_c = 2048$
64	1	$R_c = 1024$	--	---	--	---
	2	$R_c = 2048$	1	$R_c = 2048$	--	---

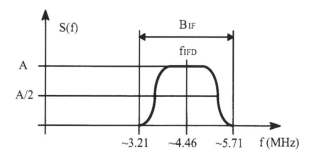

Figure 6-5. Spectrum of the CDMA signal generated by computer simulation.

The 16 bit digitized simulated waveform (including the pseudo-random stream of information symbols and the UW) is then saved into a binary file that is subsequently loaded into the memory bank of an AWG computer board. The AWG is National Instruments' PCI-5411 and is inserted into a PCI slot of the master PC. It has an 8 Msample RAM, whereby each sample is represented on 16 bit (and the overall number of stored samples must be a multiple of 8). Also the rate of digital to analog conversion (DAC) can be set to 20 or 40 Msample/s.

The AWG reads the samples in digital format from the memory at frequency $f_{AWG} = 20$ MHz. Such a value is imposed by the characteristics of the board and cannot be easily modified. Therefore, the simulation program features *interpolation* of the signal samples generated at 16.384 MHz to 're-sample' them at 20 MHz.

The number of stored digital samples per chip interval is then

$$N_c = f_{AWG} / R_c ,$$ (1.1)

and the number of stored samples per symbol interval turns out to be

$$N_s = T_s \ f_{AWG} = 2 \ f_{AWG} / R_b .$$ (1.2)

Taking into account that the RAM storage capability is 8 Msample, the maximum number of samples stored in the AWG memory amounts to

$$N_{max} = 8M / N_s = R_b / 5 .$$ (1.3)

The values of N_s and N_{max} are reported in Table 6-2, for different values of the bit rate.

Table 6-2. Values of N_{max} and N_{sps} as a funcion of R_b.

R_b (kbit/s)	N_{max}	N_s
2	400	20000
4	800	10000
8	1600	5000
16	3200	2500
32	6400	1250
64	12800	625

In order to generate a test waveform with arbitrary duration, the file containing the signal samples s_k must be read *cyclically* by the AWG. Therefore special care must be devoted to ensuring continuity of the signal at the edges of the frame. In particular, the tails of the pulses at the end of the frame must be 'wrapped around', so as to make the signal appear periodic. Considering the UW, the total number of symbols transmitted in every frame by the AWG is $N_{MEM} = N_S T_X + 24$, where $N_S T_X$ represents the number of information symbols. Also, because of the wrap around issue, the total number of symbols generated by the FORTRAN simulator is increased by 3 (Figure 6-6), both to accommodate possible signal delays up to one symbol interval, and to keep into account the tails of the chip pulses in the last symbol of the stream.

When the samples stored in the memory bank are read cyclically, the generated waveform turns actually out to be periodic, but, by carefully selecting the length and kind of the symbols pattern within each period, it can be considered as a random signal to a good approximation. This can be accomplished by using a maximal length pseudo-noise (PN) sequence as the bit stream. The PN sequences have a repetition period $L_{PN} = 2^q - 1$, with q any integer, and this implies $L_{PN} < N_{MAX}$. The parameters of the data generator polynomials are reported in Table 6-3.

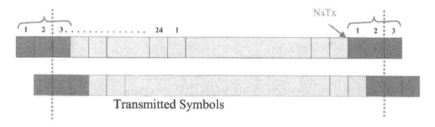

Figure 6-6. Wrap around in the generation of the waveform.

Table 6-3. Parameters of the generator polynomials.

Symbol number N_{MAX}	Code period L_{PN}	Bit rate R_b (kbit/s)	Taps	Generator polynomial of the useful channel's data (PN1)	Generator polynomial of the interfering channels' data (PN2)
400	255	2	8	8,4,3,2	8,6,5,3/8,6,5,2/8,7,6,5,2,1
800	511	4	9	9,4	9,6,4,3/9,8,5,4
1600	1023	8	10	10,3	10,8,3,2
3200	2047	16	11	11,7,3,2	11,8,5,2
6400	4095	32	12	1,6,4,1	12,9,3,2
12800	8191	64	13	13,4,3,1	13,10,9,7,5

The data stream of the I-component of the useful signal is given by the PN1 sequence, whilst the stream of the Q-component is given by the PN1 sequence shifted by half a repetition period of the sequence itself. The I-component of the generic CDMA interferer is given by the PN2 sequence with a different shift for each user. The interfering Q-component data streams are obtained by shifting of half a period the sequences of the relevant I-component streams. The number of signal samples stored in the memory of the AWG must be a multiple of 8, and in some cases this requires a modification (i.e., shortening) of the code period. Table 6-4 presents the parameters of the data generators together with the relevant AWG memory occupancy.

In the following, each period of the transmitted waveform generated by the AWG will be denoted as a *frame*, each frame starting with a UW containing 24 known symbols. The AWG outputs an analog waveform $y(t)$ which undergoes DAC aperture compensation and low pass anti-image filtering. In Figure 6-7 the interpolation filter $P(f)$ which equalizes the aperture distortion introduced by the DAC is shown inside the AWG, but in reality it is implemented off-line in the FORTRAN simulation, before the waveform actually undergoes DAC.

Table 6-4. Parameters of the data sequence generators and memory occupancy.

Symbol number N_{MAX}	Length of the code period plus UW	Length of the modified code period plus UW	Bit rate R_b (kbit/s)	Number of stored samples	Occupied memory (Mbyte)
400	255 + 24	255 + 24	2	5580000	10.64
800	511 + 24	511 + 24	4	5350000	10.2
1600	1023 + 24	1023 + 24	8	5235000	9.98
3200	2047 + 24	2046 + 24	16	5175000	9.87
6400	4095 + 24	4092 + 24	32	5145000	9.81
12800	8191 + 24	8184 +24	64	5130000	9.78

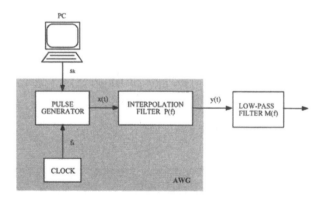

Figure 6-7 Sketch of the computer-based CDMA test signal generation.

Figure 6-8 shows the (chip time) eye diagram of a single-user CDMA signal generated during a preliminary validation run of the FORTRAN simulator. We will also show its spectrum in next sections.

After digital to analog conversion performed at the AWG output, an active mixer based upconverter brings the transmitted signal from the digital intermediate frequency f_{IFD} to the desired standard intermediate frequency

$f_{IF} = 70$ MHz. The upconverter makes use of a Local Oscillator (LO) with frequency

$$f_{LO} = f_{IF} - f_{IFD} = 65.536 \text{ MHz.} \tag{1.4}$$

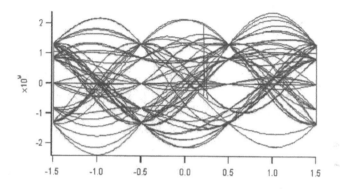

Figure 6-8. Eye diagram of a single CDMA signal ($R_b = 4$ kbit/s, $R_c = 256$ kchip/s, $L = 128$).

The image replica arising from the upconversion process is suppressed by means of an analog SAW filter with fixed bandwidth of 2.5 MHz, to take into account of the maximum signal bandwidth.

Finally, a precise amount of noise is added to the transmitted signal to reproduce the typical impairment of a satellite channel (downlink) scenario. The testbed relies on the wide band white Gaussian noise generator/adder, UFX7107 by NoiseCom™. Its output signal is almost flat in the range 0–100 MHz and its spectral power density is set with 0.1 dB accuracy.

1.3 The Master Control Program

A key role in the MUSIC testbed is played by a custom developed LabVIEW application, which control over every phase of the MUSIC experiment [MUS01].

LabVIEW is a programming environment which was originally conceived to build up applications for process control and remote measurement with a simple, quick and flexible graphical user interface. Later, it developed into a powerful environment that allows the development of general purpose programs (including control, GUI development and so on) in the form of block diagrams instead of a series of written statements as in ordinary high level languages. A wide collection of predefined libraries provides the programmer with lot of functions for data acquisition, analysis, display and storage, by means of either GPIB *(General Purpose Interface Bus)* or serial ports. Programs written in Labview are also referred to as virtual instruments (VIs) since they emulate real instrumentation, both in the appearance and in the operating mode.

The interconnection between PC and pieces of instrumentation is based on the GPIB standard developed by Hewlett Packard in two versions *(IEEE 488-1975* and the *IEEE 488.2-1987)* operating at 1 Mbyte/s. The GBIP bus is implemented also in the National Instruments devices.

The LabVIEW master control application running on the master PC of Figure 6-2 sets up the CDMA transmitter and downloads the output of the FORTRAN simulation to the AWG board. As far as signal generation is concerned, the testbed operator can set the number of active channels, the chip and bit rates, the code length and type, and the number of active users. The roll off factor of the chip pulse shaping filter can be set as well. A sample interface screen for the CDMA generator is shown in Figure 6-9. The 'advanced settings' panel shown in Figure 6-10 allows us to individually set the features proper of any interfering signal, such as signature sequence identifier and normalized delay, frequency and phase offset and power ratio with respect to the useful channel.

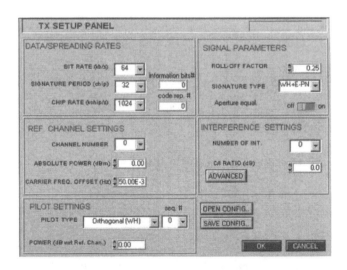

Figure 6-9. Master control program GUI for modulator parameters setting.

The Master control application also sends out all signal configuration parameters to the receiver board via an RS232 connection. Figure 6-11 shows a sample screen of the receiver configuration to be sent to the PROTEO board [Fan01], [De03a]. After receiver configuration and set up is done, a proper calibration procedure is automatically started by the Master control program to ensure that the values of the desired SNR and signal to interference ratio are correctly set. Calibration is based on a set of measurements carried out via the spectrum analyzer and controlled via GPIB by the master PC.

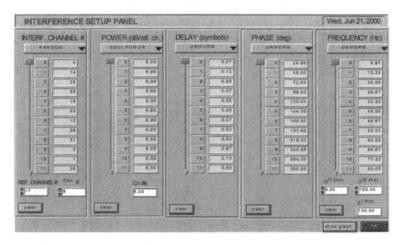

Figure 6-10. Master control program GUI interference advanced setting.

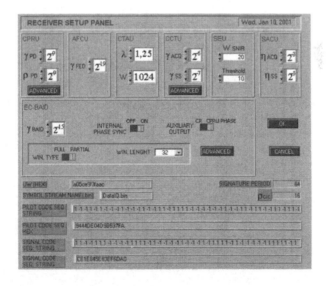

Figure 6-11. Master control program GUI for receiver parameters setting.

After set up and calibration the receiver derives the code, carrier, and framing references and starts decoding the incoming data stream, so that the Master control application can start the BER performance evaluation procedure. The DSP on the PROTEO breadboard is the unit in charge of accomplishing this task through a procedure of direct error counting. Specifically, it compares the transmitted bits (as read from the relevant file produced by the FORTRAN simulator and sent via RS232 to the receiver board) with their estimates as derived by the EC-BAID detector.

The 24 symbol UW is also used to resolve the ambiguity inherent to the process of carrier phase estimation. As described in Chapter 3, the conventional QPSK phase detector used for the MUSIC receiver suffers form a phase ambiguity of $\pi/2$, which is solved with aid of the UW as follows: once the location of the UW is found through a non-coherent algorithm the DSP performs all possible $\pi/2$-multiple counter-rotations of the symbol stream coming from the EC-BAID demodulator in the UW period, and compares the respective results with what they should be in the absence of phase error, to find out the most likely.

Eventually the BER measurements are sent back to the master PC where they are post-processed by the LabVIEW Master control application for visualization.

2.   TESTBED MONITORING AND VERIFICATION

The FPGA implementation of the MUSIC receiver was carefully tested by means of an accurate debugging procedure that aimed at demonstrating the perfect match between the outputs of the synthesized circuits and their RTL descriptions. In other words, the objective of this activity was to demonstrate that no failures occurred during the digital synthesis design flow addressed in Chapters 4 and 5. To accomplish this task the receiver was equipped with additional modules especially conceived to increase system controllability and monitoring (e.g., additional plug in boards interfaced to the CPLD headers).

In addition to allowing preliminary debugging, such monitoring resources also permitted enhanced testability and control of the receiver status during operation [MUS01].

2.1   Testbed Debugging Features

Before going into foundry with the EC-BAID ASIC, the specific detector and the general receiver design were verified by means of an especially designed HW test bench. In particular, two identical breadboards, henceforth called PROTEO-I and PROTEO-II were used. The first one (PROTEO-I) was dedicated to the implementation of the multi-rate front end, whilst the second one was temporarily used for a preliminary FPGA-based implementation of the EC-BAID detector [MUS01]. In the very first testing and debugging stage PROTEO-II, acting as the test bench breadboard, was also used to host a programmable finite state machine (FSM) pattern generator to feed PROTEO-I (under test) with the samples of an arbitrary stimulus, and to gather the related output; the latter was then compared with the expected waveform, as obtained by bit true computer simulations. The boards were connected through flat ribbon cables and the input stimuli, as well as the expected outputs, were stored in two dedicated ROM memories of the test bench board.

2.1.1 Multi-Rate Front End Verification

During verification of the multi-rate front end accommodated in the PROTEO-I board, the EABs of the test bench board (PROTEO-II) were loaded with the samples of a CDMA signal segment, as generated by the FORTRAN bit true simulator. The maximum allocable memory in a single CPLD device is 2 Kbyte, so that only 2048 8 bit input samples could be stored. Verification was carried out at the maximum chip rate $R_{c,\max} = 2048$ kchip/s, in order to strain the data paths as much as possible. Assuming the code length $L = 64$ and considering 8 samples per chip, it turns out that only 4 data symbols can be stored in a single ROM.

The most significant observable digital signals in the multi-rate front end are the In Phase (I) and Quadrature (Q) interpolator outputs. They are updated at the rate $4R_c = 8192$ MHz (every other master clock period). This feature may be exploited by introducing time multiplexing, thus both I and Q components can be alternatively compared with the expected waveforms produced by the FORTRAN bit true simulator and stored in the EABs on the test bench board.

The outcome of the front end verification is shown in Figure 6-12 and Figure 6-13, where the flag `compare_out`, output by the FSM, is reported as displayed on the oscilloscope screen and on the waveform viewer of the Synopsys VSS simulation tool, respectively. This is the result of the logical comparison between the multiplexed output of the interpolators described above, and the relevant expected waveform: a logical value '1' denotes equality. Why is it that this signal goes to 0, then to 1 again with a few intermediate spikes? The reason is that both figures present a close up of what happens at the end of the test, when the input stimulus signal out of the ROM is zeroed, and the expected signal is zeroed as well. The output signal cannot abruptly go to zero because of the presence of the front end filters, so it actually reaches zero after a transient period due to the tails of the filters response. Therefore the flag jumps again to 1 at the end of this transient period. The intermediate spikes are the result of the filter output occasionally passing through zero, and thus `compare_out` signals a 'false equality'. Nevertheless, the FSM freezes the value of `compare_out` into flag `match` (also reported in Figure 6-13) just before the beginning of the transient, not to affect the result of the test.

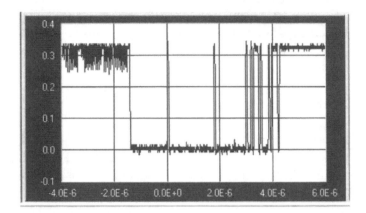

Figure 6-12. Close up of signal `compare_out` on the oscilloscope screen.

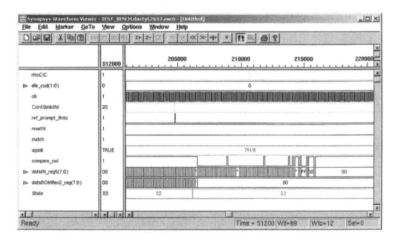

Figure 6-13. Front end verification on the Synopsys waveform viewer.

2.1.2   Synchronization Loops Verification

Once the front end was successfully verified, the acquisition and tracking loops, implemented by the CTAU and CCTU (with their ancillary SAC unit) were also verified. In order to make the test long enough to correctly stimulate these units, the code length was set to $L = 32$, and the chip rate was the maximum allowed, namely, $R_c = R_{c,\max} = 2048$ kchip/s. The sync loops are directly fed with the interpolated signal at rate $4R_c$, therefore the ROM memory storing the input stimulus spanned 8 symbol periods. After the CTAU has performed coarse code acquisition, the CCTU loop initiate chip timing tracking, and starts producing the two control signals `fract_del` and `int_del`, corresponding to parameters L_k and μ_k, respectively (see Chapter 3).

Two signals can be selected, according to the value of the configuration bit `test_sel`, to be compared for testing purpose in the test bench board, namely, either the 7 bit AGC gain `rho_AGC`, or the 7 bit composite signal obtained by concatenation of `fract_del` (most significant 5 bits) and `int_delay` (least significant two bits).

The FSM compares the selected signal (one sample per symbol period) with the waveforms predicted by the Fortran bit true simulator. To achieve further testing flexibility, and stress the CCTU loop, a 2 bit parameter `test_mode` was also introduced. The permitted values are '00' (mode 0), '01' (mode 1) and '11' (mode –1). In mode 0 the FSM cyclically reads the ROM input memory; in mode $+1$ (-1) the first sample of the memory is read twice (not read) every 32 full reading cycle, so that the CCTU synchronization loop has to track a $T_c/8$ delay (advance) every $32 \cdot 8 = 256$ symbols.

The digital AGC gain is reported in Figure 6-14 and Figure 6-15 as retrieved from the hardware test bench and as expected from software simulations, respectively, both in mode 0. The test window is 2048 symbol periods long, which is 32 ms at the symbol rate $R_s = 64$ Ksymb/s. The initial low level in the AGC transient is owed to the initial receiver reset, whilst the intermediate one before the testing start, is caused by waiting for signal synchronization from CTAU. Finally, after test completion the AGC gain straight grows toward saturation, because no signal to be regulated is presented at the SAC input.

Figure 6-16 and 6-17 depict `fract_del` in mode 1 as it is displayed on the scope, as well as derived from SW simulations. In test mode 1 one input sample is read twice every 256 symbols, which means every $256/R_s = 4$ ms, therefore the CCTU loop has to follow the signal delay of $T_c/4$ every 8 ms, which occurs four times in the considered observation window.

2.1.3   EC-BAID Verification

Once synthesized into the FPGAs of PROTEO-II, the EC-BAID circuit was tested separately from the MUSIC receiver to ensure perfect matching between the desired 'bit true' model and the FPGA implementation. The PROTEO-I breadboard was loaded with the hardware test bench conceived to generate the chip rate input samples and to retrieve the EC-BAID outputs.

Proper software running on the DSP was in charge of verifying the match between the hardware outputs and the expected values. Particularly, 1024 chip long test vectors were generated by the FORTRAN software and stored in the PROTEO-I RAM to be circularly sent to the PROTEO-II (EC-BAID) breadboard, whereas the DSP was in charge of storing and comparing up to 16304 output values. Several tests, reported in Table 6-5, were performed to validate the implementation of the various EC-BAID blocks, and all of them were successfully run. Particularly, in the last two tests some timing re-alignment operations (which are likely to happen in a realistic environment) were also successfully emulated.

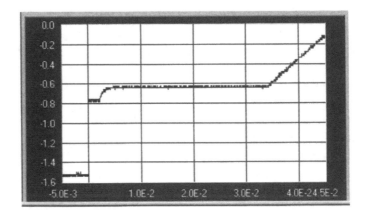

Figure 6-14. AGC gain `rho_AGC` transient as displayed on the scope (mode 0).

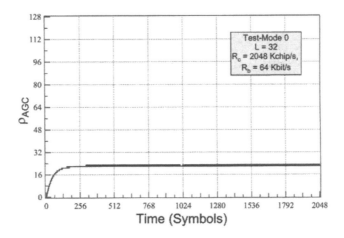

Figure 6-15. AGC gain `rho_AGC` transient retrieved from Fortran simulations (mode 0).

According to the Static Timing Analysis results reported in Chapter 4, this EC-BAID FPGA implementation properly worked at all the chip rates up to 512 kchip/s; the same preliminary tests also confirmed that the FPGA circuit did not work at R_c = 2048 kchip/s, whereas it actually worked at R_c =1024 kchip/s in spite of the (conservative) theoretical timing analysis. The chip rate of 512 kchip/s was then chosen for all functional tests with both the MUSIC receiver and EC-BAID breadboards reported in Table 6-5.

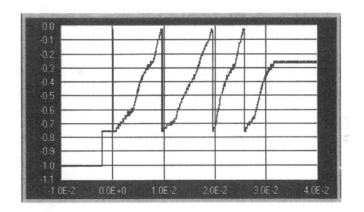

Figure 6-16. CCTU `fract_del` AS displayed on the scope (mode 1).

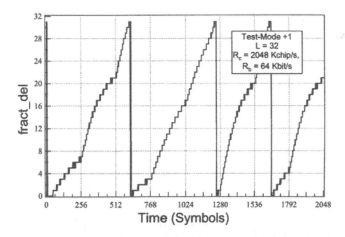

Figure 6-17. CCTU `fract_del` derived by Fortran simulations (mode 1).

2.2 Debugging the MUSIC Receiver

Once the PROTEO-II board with the FPGA implementation of the EC-BAID is tested in a stand-alone mode joint testing of the two boards with both the EC-BAID and the receiver was started. In particular, testing was mainly focused on the temporal and spectral representation of the key internal signals of the receiver, during its normal operation. The oscilloscope, the spectrum analyzer, and the visualizer provided with the Boxview DSP software running on the master PC were our analysis tools [De03a], [De03b], [MUS01].

Table 6-5. Hardware EC-BAID tests.

Test number	EC-BAID alg. ON	CPRU ON	Phase to be recovered	Timing corrections	Tested output
1			0°		AGC gain
2			0°		CR symbols
3			0°		CR accumulator
4			0°		BAID symbols
5		✓	0°		BAID symbols
6		✓	45°		BAID symbols
7	✓		45°		BAID symbols
8	✓	✓	45°		BAID symbols
9	✓	✓	45°	✓	AGC gain
10	✓	✓	45°	✓	BAID symbols

As mentioned in the previous sections, the PROTEO breadboards were also equipped for improved testability with an external plug in DAC board, specifically designed for the project, and inclusive of anti-image analog filters (not shown in Figure 6-3). As mentioned above, this board allowed the monitoring in the time and frequency domains of the key receiver signals such as the CDMA baseband converted signal, the CTAU and CCTU lock detectors, etc., as mentioned above. Selection of the desired signal was done at run time by re-configuring appropriate multiplexers implemented on both CPLDs via the DSP.

Flex-I is equipped with a four-input multiplexer, which makes it possible to select one signal out of IF input, In Phase (I) baseband signal, I CIC output and I CMF output. The 2 bit control signal is passed to the DSP interface as a configuration parameter and can be changed on the fly. The addressed signal is finally sent to flex-II, whereupon the DAC card is mounted.

The reset signal reset_n is treated the same way as other configuration parameters, and is supported by a register located into the DSP interface of flex-I. This means the DSP software is in charge of resetting the receiver at its start up, completing the configuration parameters setting, and finally removing the reset.

One other 16 input multiplexer is located on flex-II. At its input we find (in addition to others) fract_del and int_del (the synchronization loop outputs, L_k, and μ_k respectively), whilst `Rho_AGC` (the AGC gain output by the SAC unit), `CRoutP` (the internal accumulator status of the AGC correlator), and finally `Mux_Flex1` (the signal selected by the multiplexer in flex-I). As for flex-I, the 4 bit multiplexer control signal is written via DSP interface and can be changed on the fly; the selected output is directly passed to the DAC board for visualization and can be read by the DSP Interface. In addition to `Mux_Flex1`, the DSP interface is directly connected to several other signals, such as the abovementioned AGC gain `rho_AGC`, the CCTU control signal `fract_del`, or `Bout_BAID` representing the concatenation on 8 bits of the two symbol rate I/Q 4 bit 'soft' outputs of the EC-BAID. The DSP interface also outputs an interrupt request towards the DSP, based on the symbol strobe output by the EC-BAID demodulator. Such interrupt is used to launch a DSP primitive implementing BER measurement.

As an example of the monitoring facilities described above, Figure *6-18* depicts the spectrum of the IF input signal, as seen on the analyzer display and with $R_b = 64$ kbit/s, $R_c = 2048$ kchip/s ($L = 64$) and $E_b / N_0 = 10$ dB.

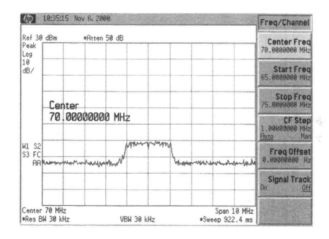

Figure 6-18. Power spectrum of the IF CDMA signal with $N = L = 64$ users.

We show now in Figures 6-19 to 6-22 a set of signal spectra obtained in a typical downlink configuration with WH synchronous channelization codes (code number 5 for the useful channel) with E-Gold scrambling code, and a non-orthogonal pilot channel (WH code 0) for synchronization and power control (not implemented). Actually, only one reference user and pilot channel have been considered in following Figures 6-19 to 6-22. The code length L is 64, the chip rate R_c is 1024 kchip/s (i.e., the input signal is oversampled by a factor 16) and the bit rate R_b is 32 kbit/s. Figure 6-19 shows the (spectrum of the) received signal at the DAC output, with the useful components centered at $f_{IFD} = 4.464$ MHz, and their spectral replica, generated by pass-band sampling at the master clock frequency $f_{clk} = 16.384$ MHz, at $f_{IF}' = 11.920$ MHz.

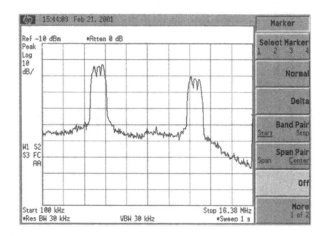

Figure 6-19. IF received digital signal with $R_c = 1024$ Mchip/s. Spectral replica is centered upon the frequency of 11.92 MHz.

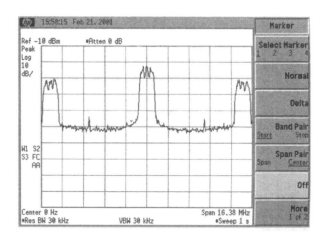

Figure 6-20. Baseband converted signal at the input of the CIC filter.

Figure 6-20 shows the same signal (I component) after baseband conversion at the output of the DCO. The sampling rate is $f_{clk} = 16.384$ MHz (just like at the ADC output), so that the two spectral replicas are located at the frequencies $\pm(f_{clk} - 2f_{IFD}) = \pm7.456$ MHz and $\pm2f_{IFD} = \pm8.928$ MHz (not shown). The spectral spikes located at $\pm f_{IFD} = \pm 4.464$ MHz are just carrier residues.

Figure 6-21 shows the decimated signal at the output of the CIC filter, where the sampling rate is $4R_c = 4.096$ MHz (decimation factor $\rho = 4$). The CIC decimator also performs cancellation of the unwanted signal replicas at $\pm (f_{clk} - 2f_{IFD})$ and $\pm 2f_{IFD}$. No in band distortion is seen, whilst images at $\pm 4R_c$ are still present.

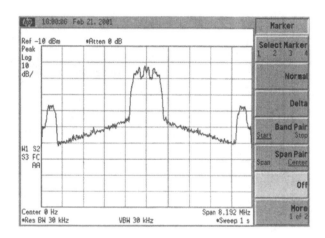

Figure 6-21. Decimated signal at the CIC output.

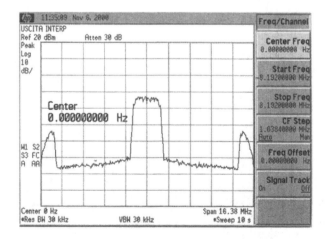

Figure 6-22. Spectrum of the CMF output signal (I component).

Figure 6-22 (relative again to $R_b = 64$ kbit/s, $R_c = 2048$ kchip/s ($L = 64$) and in the same noisy environment of $E_b / N_0 = 10$ dB as Figure 6-18) shows the power spectrum of the I component at the output of the CMF. The spectral image is located at $f_d = 4R_c = 8.172$ MHz and the out of band noise power reduction performed by the digital front end as a whole is apparent if we compare this spectrum with the one in Figure 6-18.

Time domain signals can be displayed either via a digital scope connected to the plug in DAC board, or via the 'virtual scope' provided by the BoxView tool. In this context Figure 6-23 sketches the I interpolator output when the (non-orthogonal) pilot is the only active channel[1].

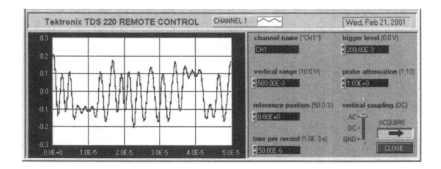

Figure 6-23. Interpolator output signal, useful channel plus pilot with P/C = 30 dB.

Figure 6-24. AGC gain acquisition transient.

Figure 6-24 shows the acquisition transient of the AGC gain signal, as displayed on the visualization tool of the BoxView DSP software. The signal is read at symbol time by means of the DSP interface, and the DSP stores its value in the data memory, so as it can be read by the Master PC for visualization. The observation window time amounts at about to 250 symbol inter-

---

[1] As the generation software did not allow the transmission without the reference channel, we emulated the pilot only condition by setting $P/C = 30$ dB.

vals. The agreement with previously produced bit true simulation results is excellent.

Figure 6-25 displays the contents of the internal accumulator of the pilot-channel correlator in the SAC unit, when $R_b = 32$ kbit/s, $R_c = 1024$ kchip/s ($L = 64$) and no noise is affecting the transmission. The resulting waveform is a periodic ramp (the pilot channel is unmodulated), whose period equals the pilot code repetition length (i.e., the symbol period interval) $T_s = 1/R_s \approx 61$ µs.

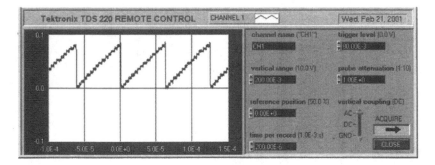

Figure 6-25. Internal status of the I pilot correlator of the digital AGC.

Figure 6-26 shows the time evolution of signal `fract_del` generated by the CCTU and controlling the re-sampling epoch input to the linear interpolator unit. This signal updated at symbol time is a ramp as the result of the (constant) clock frequency shift between transmitter and receiver, which causes the optimum chip sampling epoch to drift uniformly.

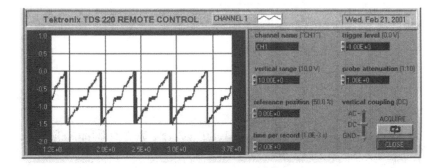

Figure 6-26. CCTU fract_del signal.

Similar debugging features were added to the FPGA implementation of the EC-BAID detector; they rely on the configuration of two control signals in order to select a proper output for the PROTEO-II breadboard. Selection is made according to Tables 6-6 and 6-7.

Table 6-6. Auxiliary output configuration.

Test_sel (2 bits)	Auxiliary output (8 bits)
00	CR output symbols (4 + 4 bits)
01	CPRU (Carrier Phase Recovery Unit) phase
10	AGC level
11	Norm of the x^e vector

Table 6-7. PROTEO-II output configuration.

Swap_sel (1 bit)	PROTEO-II output (8 bits)
0	EC-BAID output symbols (4 + 4 bits)
1	EC-BAID / Auxiliary outputs multiplexed

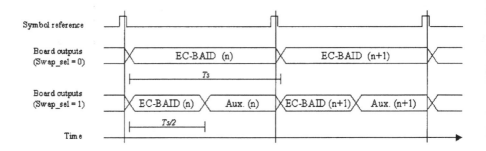

Figure 6-27. EC-BAID internal outputs timing diagram.

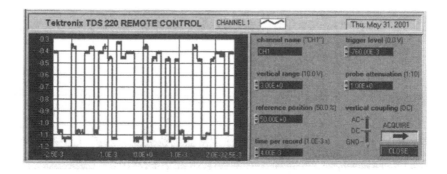

Figure 6-28. EC-BAID outputs with no interferers and Eb/No → ∞.

The configuration parameter Test_sel selects the kind of auxiliary output to be provided, while Swap_sel determines the behavior of the breadboard outputs: when Swap_sel is '0' the I/Q EC-BAID outputs are sent on the bus, whilst with `Swap_sel` equal to '1' the I output of the EC-BAID is multiplexed with the one of the auxiliary interference-mitigating correlator, according to the timing diagram in Figure 6-27.

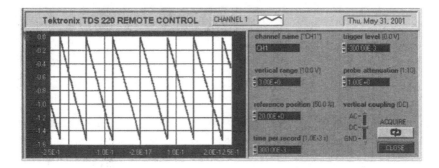

Figure 6-29. Phase acquisition with frequency offset.

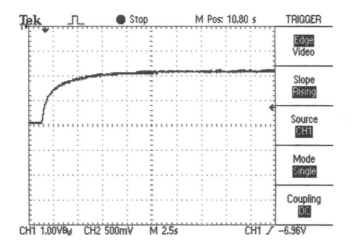

Figure 6-30. |x| transient.

All of the four observable signals as in Table 6-7 were used to test the match between our FORTRAN bit true model and VHDL design, as detailed in Chapters 4 and 5. As an example, Figure 6-28 shows the EC-BAID soft output for the case of an ideal transmission with no interferers and E_b / N_0 approaching infinite. Figure 6-29 shows a typical phase acquisition curve when the CPRU operates with a residual frequency error at the EC-BAID input. Finally, Figure 6-30 is a 'summary' of the acquisition phase of the

EC-BAID since it represents the norm of the adaptive vector $\mathbf{x}^e$ that gives interference mitigating capability (see Chapter 3). This quantity allows to detect whether the EC-BAID is in its steady state or not, and was also used to detect and correct timing violations in the critical path of our FPGA design. Timing violations look like sharp spikes in the curve of $|\mathbf{x}|$ that of course are not present in Figure 6-30 since it represents a sample of correct operation.

3.   OVERALL RECEIVER PERFORMANCE

Once the debugging phase was complete a set of receiver performance measurements (mainly in the form of BER curves) was planned and carried out. The goal was to cover as extensively as possible the transmission conditions and system configurations listed in the project specifications. According to Table 6-1, the code length L was varied into the range 32–128, the chip rate R_c spanned the interval 128 kchip/s to 2048 kchip/s and consequently the symbol rate R_s ranged from 1 to 32 ksymb/s.

All of the numerical results presented hereafter were derived in the presence of a synchronous non-orthogonal E-Gold pilot signal code division multiplexed with the useful traffic channels as discussed in Section 3 of Chapter 2. The pilot to useful carrier power ratio P/C was set to 6 dB as a good trade off between interference level owed to residual cross correlation and sync accuracy provided by pilot aided operations. We also defined as *mild*, *medium* and *heavy* load conditions those corresponding to a total number of active channels (encompassing the useful, the pilot and the interfering ones) $N = L/4$, $N = L/2$ and $N = 3L/4$, respectively. The interfering channels are equi-powered and the 'useful carrier to single-interferer' power ratio is set to $C/I = 0$ dB. The update step size of the EC-BAID algorithm, introduced in Chapter 3, was $\gamma_{BAID} = 2^{-13} = 1.44 \times 10^{-4}$ for mild and medium loading and $\gamma_{BAID} = 2^{-15} = 3.05 \times 10^{-5}$ for heavy loading. The leak factor, also defined in Chapter 3 was set to the optimum value $\gamma_{leak} = 2^{-3} = 0.125$.

The benchmarks for our experimental results were the corresponding BER curves obtained after floating point and/or bit true software simulations. In particular, we resorted to long simulations to increase the BER estimation accuracy as much as possible. Two different configurations were selected: 200K symbols long transient and 100K symbol of observation for BER estimation when $\gamma_{BAID} = 2^{-13}$, 250K of transient and 50K useful symbols when $\gamma_{BAID} = 2^{-15}$. Before comparing HW measurement results with simulations, we also performed fine tuning of the different receiver parameters (leakage factor, chip timing loop bandwidth, etc.) by direct observation of the HW behavior, and we re-run our simulations accordingly. Table 6.8 reports the main setting as determined through this activity.

Table 6-8. Optimum values for receiver setting parameters.

Parameter	Optimum experimental value
$\gamma_{CCTU, acq}$	2^{-7}
$\gamma_{CCTU, ss}$	2^{-7}
$\gamma_{AGC, acq}$	2^{-2}
$\gamma_{AGC, ss}$	2^{-4}
γ_{AFC}	2^{-15}
γ_{CPRU}	2^{-9}
ρ_{CPRU}	2^{-9}
γ_{LEAK}	2^{-3}
$\gamma_{AGC\ BAID}$	2^{-4}

It is time now to present some of our most significant experimental BER results, excerpted from a wider collection reported in [MUS01]. Concerning notation, in all of the following charts the label 'sw' and white marks denote numerical results obtained by computer simulation of the whole system (including all the sync loops) carried out with floating point precision, whilst the label 'hw' and colored marks refer to measured results. Figure 6-31 compares SW and HW EC-BAID's BER performance for $L = 64$ and $R_c = 512$ kchip/s in the absence of MAI (apart from the pilot which is always assumed active).

Figures 6-32 and 6-33 present the BER curves for $L = 32$ and $R_c = 512$ kchip/s, but in the case of medium and heavy load conditions, respectively. Simulated BERs of the conventional CR are also reported for the sake of comparison. Finally, Figures 6-34 and 6-35 compare the simulated and measured BER performance for $L = 128$ and mild loading, with $R_c = 1024$ kchip/s and $R_c = 2048$ kchip/s, respectively. These results clearly show that the implementation loss of the whole receiver is about 1.0 to 1.5 dB at the target BER of 10^{-3} for the selected configurations. This figure includes *all* losses experienced by the system: signal generation, distortion owed to analog IF processing, signal quantization, synchronization loops, etc.. In particular, the TX and RX clocks were *not* locked as is often done in back to back laboratory breadboard evaluations, so the impairment owed to TX/RX clock misalignment is also taken into account.

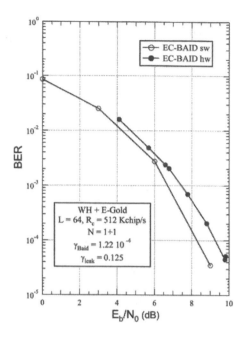

Figure 6-31. Experimental BER performance —— see chart inset for parameters values.

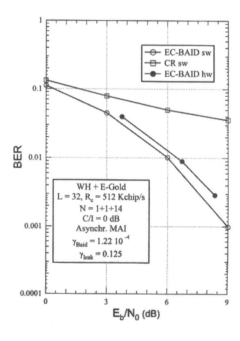

Figure 6-32. Experimental BER performance —— see chart inset for parameters values.

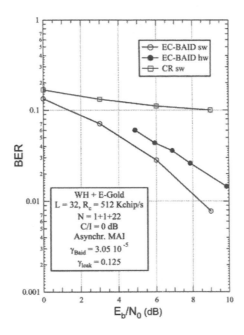

Figure 6-33. Experimental BER performance —— see chart inset for parameters values.

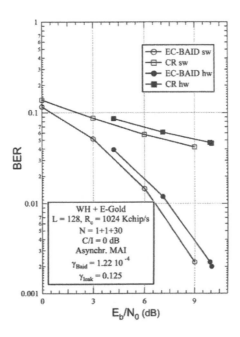

Figure 6-34. Experimental BER performance —— see chart inset for parameters values.

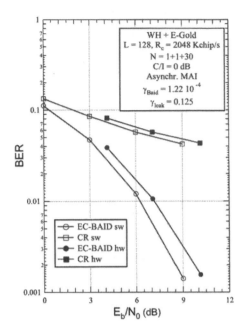

Figure 6-35. Experimental BER performance —— see chart inset for parameters values.

Chapter 7

CONCLUSION?

No, the question mark in the title of this Chapter is not a typo. In the few pages to follow we will try to convince the reader that the issue of good, efficient design of a wireless terminal with non-conventional signal processing functions is far from being concluded. To accomplish this, we will first summarize what, in our opinion, are the main outcomes of the MUSIC project. And then we will outline a few questions that are worth being pursued in the future. We do hope that, in some lab, under the cover of IPs and industrial secret, some researcher has already started pursuing them...

1. SUMMARY OF PROJECT ACHIEVEMENTS

At the moment, no one doubts about CDMA being a key technology for the successful implementation and deployment of present-time 3G and (much likely) future 4G wireless communication networks. The MUSIC project, supported by the ESA Technology Research Programme (TRP), has successfully demonstrated that advanced digital signal processing techniques are effective in mitigating CDMA interference, thus contributing to increase the capacity and/or quality of service of a wireless communication network (be it satellite or terrestrial).

As the reader should have clear by now, the low-complexity interference-mitigating solution investigated and developed in the project is particularly suited for being implemented in mobile terminals. In addition to demonstrating a good agreement of measurements with theoretical and simulation results, the project has also demonstrated the possibility to integrate advanced CDMA interference-mitigation techniques into a single ASIC device. In particular, the design flow adopted when implementing ancillary functions on FPGAs allows an easy re-use of the resulting architecture to come to an overall integration of the receiver into a single ASIC with modest complexity and power consumption. Of course, interference mitigation is not the sole advanced DSP feature that has to be incorporated into an advanced wireless terminal. Channel coding, audio/video/image compression, challenge the designer as well. The issues to be faced for efficient System-on-a-Chip (SoC) design are still the fundamental ones: complexity and power. These two factors have to be carefully traded-off for pure performance to come to a final efficient design. The project continually faced such an issue, and we hope that the book succeeded in clearly showing this to the reader.

A further fundamental achievement lies in the area of *methodology* rather than in the domain of "pure" state-of-the-art results. The project team is convinced of having attained the right attitude for close cooperation between system-level and HW-level designers, which in a word leads to efficient co-design. At the end of the project all communication engineers started "thinking HW", in the sense that they could re-formulate their algorithms from the very start in order to make them easier to implement by the VLSI/chip architects. And the latter could at the end of the project suggest many non-trivial modifications to the system-level DSP algorithms to make them more efficient form the "pure performance" standpoint (i.e., they learnt "thinking DSP"). This is actually the Holy Graal of every communication terminals design team.

2.   PERSPECTIVES

Although the role of satellites in 3G system is currently still being debated, it seems reasonable to assume that the satellite network can integrate with IMT terrestrial networks to carry out two fundamental functions: *i*) enhancing the modest broadcasting capabilities of the terrestrial network, and *ii*) acting as a gap-filler in poorly covered areas. Thus, the design and low-cost implementation of dual-mode terminals operating on similar carrier frequencies appears a mandatory issue. Just like mandatory appear further studies and experimentation on the performance of interference-mitigation in a mixed satellite/terrestrial environment for the reasons above. This is the approach pursued by ESA, which is about to complete the development of a comprehensive 3G W-CDMA satellite UMTS (S-UMTS) test-bed that will allow to fully characterize the EC-BAID performance in the forward link of a multi-beam multi-satellite (mobile) environment [Cai99]. To this respect, more is still to be done from the theoretical as well as the experimental point of view to assess the performance of adaptive interference-mitigation on a terrestrial mobile radio channel which is mainly plagued by *frequency selective* fading effects.

One other issue that was not touched upon in this project is the interplay between IMD and powerful channel coding techniques such as Turbo

[Ber96] and Low-Density Parity Check (LDPC) codes [Mac99]. As is known, the two are both based on powerful iterative decoding techniques that allow to attain unprecedented BER performances, remarkably close to the celebrated Shannon's bound. We see that a few misconceptions are building up around this issue, notably that IMD is useless when Turbo or LDPC coding is used. We do not believe so, and we hope to see soon a book on this topic ...

A third issue that in our opinion is to be tackled soon is the applicability of CDMA with IMD techniques to decentralized, infrastructureless, ad-hoc networks [Jab02].A decentralized wireless network has no "reference" nodes (as the base stations in a cellular system) since any node can at the same time act as terminal or intermediate with routing functions: this enables the wireless network to establish multi-hop communication links, just as is commonplace in fixed networks. Such architecture lets one to envisage a communication scenario characterized by low-to-medium capacity and very low cost, with easy and flexible access. Some applications of such a scenario come immediately to our mind as for instance a "private citizen network" that develops with no infrastructures to connect a group of users within a metropolitan area (e.g., students within a University campus, workers in a big plant or in an airport etc.). A more ambitious scenario might also be a ubiquitous vehicular network whose nodes are standard radio-communication terminals aboard cars. This would give each car a certain communication capacity to handle not only automotive-related information (such as traffic control, weather, emergency etc.), but low-rate multimedia communications as well (audio, image, Internet browsing etc.). Is CDMA suited to this picture? Is IMD a good feature to increase capacity in such scenario? Will ad-hoc networking be the winning paradigm for 4G systems? Will wireless ad-hoc terminals with advanced DSP and routing functions be implementable as a SoC? Trying to answer questions like these, and succeeding in doing so, is what we call, "the pleasure of doing good research". For a better living, everyone should experience such a thrill. Even once in a lifetime might be enough ...

References

[Ada97]   F. Adachi, M. Sawahashi, K. Okawa, "Tree-Structured Generation of Orthogo-
           nal Spreading Codes with Different Lengths for Forward Link of DS CDMA
           Mobile Radio", IEE Electronics Letters, Vol. 33, No. 1, 2nd January 1997,
           pp. 27-28.
[Ada98]   F. Adachi, M. Sawahashi, H. Suda, "Wideband DS-CDMA for Next-Generation
           Mobile Communications Systems", IEEE Communications Magazine, Septem-
           ber 1998, pp. 56-69.
[Ahm75]   N. Ahmed, K.R. Rao, Orthogonal Transforms for Digital Signal Processing,
           New York: Springer-Verlag, 1975.
[Alb89]   T. Alberty, V. Hespelt, "A New Pattern Jitter-Free Frequency Error Detector",
           IEEE Transactions on Communications, Vol. 37, No. 2, February 1989.
[aptix]   http://www.aptix.com
[arc]   http://www.arc.com/
[arm]   http://www.arm.com
[Ber96]   C. Berrou, A. Glavieux, "Near Optimum Error Correcting Coding and Decod-
           ing: Turbo Codes", IEEE Transactions on Communications, Vol. 44, No. 10,
           pp 1261-1271, October 1996.
[Bha02]   H. Bhatnagar, Advanced ASIC Chip Synthesis, Kluwer Academic Publishers,
           2002.
[Bou02]   D. Boudreau, G. Caire, Corazza G.E., R. De Gaudenzi, G. Gallinaro, M. Luglio,
           R. Lyons, J. Romero-Garcia, A. Vernucci, H. Widmer, "Wideband-CDMA for
           the UMTS/IMT-2000 Satellite Component", IEEE Transactions on Vehicular
           Technology, Vol. 51, No. 2, March 2002,. pp. 306-331.
[caden]   http://www.cadence.com
[Cai99]   G. Caire, R. De Gaudenzi, G. Gallinaro, R. Lyons, M. Luglio, M. Ruggieri, A.
           Vernucci, H. Widmer, "ESA Satellite Wideband CDMA Radio Transmission
           Technology for the IMT-2000/UMTS Satellite Component: Features & Per-
           formance", Proc. IEEE GLOBECOM '99, Rio De Janeiro, Brazil, 5-9 Decem-
           ber 1999.
[celox]   http://www.celoxica.com

[Chi92]   S. Chia, "The Universal Mobile Telecommunication System", IEEE Communications Magazine, December 1992, pp. 54-62.

[Cla00]   T. Claasen, "First time right silicon but to the right specification", IEEE DAC 2000, Los Angeles, Keynote, CA, USA, 5-9 June 2000.

[Cla93]   F. Classen, H. Meyr, P. Sehier, "Maximum Likelihood Open Loop Carrier Synchronization for Digital Radio", In the Proceedings of the IEEE ICC 93, Geneva, Switzerland, May 1993.

[cowar]   http://www.coware.com

[Dah98]   E. Dahlman, B. Gudmundson, M. Nilsson, J. Sköld, "UMTS/IMT-2000 Based on Wideband CDMA", IEEE Communications Magazine, September 1998, pp. 70-80.

[DAn94]   N.A. D'Andrea, U. Mengali ,"Noise Performance of Two Frequency-Error Detectors Derived from Maximum Likelihood Estimation Methods", IEEE Transactions on Communications, Vol. 41, No. 2, February 1994.

[De03a]   R. De Gaudenzi, L. Fanucci, F. Giannetti, E. Letta, M. Luise, M. Rovini, "Design, Implementation and Verification through a Real-Time Test-Bed of a Multi-Rate CDMA Adaptive Interference Mitigation Receiver for Satellite Communication", IJSCN, International Journal on Satellite Communications and Networking, Vol. 21, Special Issue "Interference Suppression Techniques for Satellite Systems", March 2003, Pages 39-64.

[De03b]   R. De Gaudenzi, L. Fanucci, F. Giannetti, M. Luise, M. Rovini "Satellite Mobile Communications Spread-Spectrum Receiver", IEEE Aerospace and Electronic Systems Magazine, Vol. 18, Issue 8, August 2003, pp. 23-30.

[De98a]   R. De Gaudenzi, F. Giannetti, M. Luise, "Design of a Low-Complexity Adaptive Interference-Mitigating Detector for DS/SS Receivers in CDMA Radio Networks", IEEE Transactions on Communications, Vol. 46, No. 1, January 1998, pp. 125-134.

[De98b]   R. De Gaudenzi, F. Giannetti, M. Luise, "Signal Synchronization for Direct-Sequence Code-Division Multiple Access Radio Modems", European Transactions on Telecommunications, Vol. 9, No. 1, January/February 1998, pp. 73-89.

[De98c]   R. De Gaudenzi, F. Giannetti, "DS-CDMA Satellite Diversity Reception for Personal Satellite Communication: Satellite-to-Mobile Link Performance Analysis", IEEE Transactions on Vehicular Technology, Vol. 47, No. 2, May 1998, pp. 658-672.

[De98d]   R. De Gaudenzi, F. Giannetti, M. Luise, "Signal Recognition and Signature Code Acquisition in CDMA Mobile Packet Communications", IEEE Transactions on Vehicular Technology, Vol. 47, No. 1, February 1998, pp. 196-208.

[De98e]   R. De Gaudenzi, L. Fanucci, F. Giannetti, M. Luise, "VLSI Implementation of a Signal Recognition and Code Acquisition Algorithm for CDMA Packet Receivers", IEEE Journal on Selected Areas in Communications, Vol. 16, No. 9, December 1998, pp. 1796-1808.

[DeG91]   R. De Gaudenzi, C. Elia, R. Viola, "Band-Limited Quasi-Synchronous CDMA: A Novel Satellite Access Technique for Mobile and Personal Communication Systems", IEEE Journal on Selected Areas in Communications, Vol. 10, February 1991, pp. 328-343.

[DeG93]   R. De Gaudenzi, M. Luise, R. Viola, "A Digital Chip Timing Recovery Loop for Band-Limited Direct-Sequence Spread-Spectrum Signals", IEEE Transactions on Communications, Vol. 41, No. 11, November 1993.

[DeG96]   R. De Gaudenzi, F. Giannetti, M. Luise, "Advances in Satellite CDMA Transmission for Mobile and Personal Communications", Proceedings of the IEEE, Vol. 84, No. 1, January 1996, pp. 18-39.

[DeM94]   G. De Micheli, Synthesis and Optimization of Digital Circuits, Mc Graw-Hill, 1994

[Din98]   E.H. Dinan, B. Jabbari, "Spreading Codes for Direct Sequence CDMA and Wideband CDMA Cellular Networks", IEEE Communications Magazine, September 1998, pp. 48-54.

[Dix94]   R.C. Dixon, Spread Spectrum Systems with Commercial Applications, New York: Wiley Interscience, 1994.

[Due95]   A. Duel-Hallen, J. Holtzman, Z. Zvonar, "Multiuser Detection for CDMA Systems", IEEE Personal Communications, April 1995, pp. 46-58.

[Fan01]   L. Fanucci, E. Letta, M. Rovini, G. Colleoni, R. De Gaudenzi, F. Giannetti, M. Luise, "A Multi-Rate Real-Time Test-Bed for Low-Complexity Adaptive Interference Mitigation in CDMA Transmission", International Workshop on Digital Signal Processing Techniques for Space Communications, DSP 2001, 1-3 October 2001, Lisboa, Portugal.

[Fan02]   L. Fanucci, M. Rovini, "A Low-Complexity and High-Resolution Algorithm to Estimate the Magnitude of Complex Number", IEICE Transactions on Fundamentals, Vol. E85-A, No. 7, July 2002, pp. 1766-1769.

[Fer99]   A. Ferrari, A. Sangiovanni-Vincentelli, "System Design: Traditional Concepts and New Paradigms", Proceeding of the IEEE ICCD 1999, pp. 2-12, Austin, Texas, USA, 10-13 October 1999.

[Fon96]   M.-H. Fong, V.K. Bhargava, Q. Wang, "Concatenated Orthogonal/PN Spreading Sequences and Their Application to Cellular DS-CDMA Systems with Integrated Traffic", IEEE Journal on Selected Areas in Communications, Vol. 14, No. 3, April 1996, pp. 547-558.

[Gar88]   F.M. Gardner, "Demodulator Reference Recovery Techniques Suited for Digital Implementation", Final Report, ESTEC Contract No. 6847/86/NML/DG, ESA/ESTEC, The Netherlands, August 1988.

[Gia97]   F. Giannetti, "Capacity Evaluation of a Cellular CDMA System Operating in the 63-64 GHz Band", IEEE Transactions on Vehicular Technology, Vol. 46, No. 1, February 1997, pp. 55-64.

[Gil91]   K.S. Gilhousen, I.M. Jacobs, R. Padovani, A.J. Viterbi, L.A. Weaver, C.E. Wheatley, "On the Capacity of a Cellular CDMA System", IEEE Transactions on Vehicular Technology, Vol. 40, No. 5, May 1991, pp. 303-312.

[Gilch]   C.E. Gilchriest, "Signal-to-Noise Monitoring", JPL Space Programs Summary 37-27, Vol. IV, pp. 169-184.

[Gol67]   R. Gold, "Optimal Binary Sequences for Spread Spectrum Multiplexing", IEEE Transactions on Information Theory, Vol. 13, October 1967, pp. 619-621.

[Gol68]   R. Gold, "Maximal Recursive Sequences with 3-valued Recursive Cross-Correlation Functions", IEEE Transactions on Information Theory, January 1968, pp. 154-156.

[Gra02]   M. Gray, "Platforms: a way to get aboard the express train", Proc. of SAME 2002, Sophia Antipolis, France, 9-10 October 2002.

[Har97]   "HSP50110 Digital Quadrature Tuner", data sheet, Harris Semiconductors, January 1997.

[Hau01]   J. Hausner, "Integrated Circuits for Next Generation Wireless System", Proceedings of the ESSCIRC 2001, Villach, Austria, 18-20 September 2001.

[hcmos]   HCMOS8D 0.18μm Standard Cells Family, CB65000 Series, STMicroelectron-ics, data sheet, March 2002.

[hitac]   http://www.hitachi.com

[Hog81]   E.B. Hogenauer, "An Economical Class of Digital Filters for Decimation and Interpolation", IEEE Transactions on Acoustics, Speech, and Signal Processing, April 1991, pp. 155-162.

[Hon95]   M. Honig, U. Madhow, S. Verdù, "Blind Adaptive Multiuser Detection", IEEE Transactions on Information Theory, Vol. 41, No. 4, July 1995, pp. 944-960.

[Hon98]   M.L. Honig, "Review of Multiuser Detection and Interference Suppression Techniques for Satellite DS-CDMA", ESA-ESTEC Report, ESA/ESTEC, The Netherlands, May 1998.

[ibm]   http://www.ibm.com

[Jab02]   B. Jabbari, "Ad-hoc/Multi-hop Networking for Wireless Internet", European Wireless 2002, Florence, Italy, Feb. 2002.

[Kas68]   T. Kasami, "Weight Distribution of Bose-Chaudhuri-Hocquenghem Codes", in Combinatorial Mathematics and Its Applications, University of North Carolina Press, Chapel Hill, NC, 1968. Also reprinted in: E.R. Berlekamp ed., Key Papers in the Development of Coding Theory, IEEE Press, New York 1974.

[Kni98]   D. N. Knisely, S. Kumar, S. Laha, S. Nanda , "Evolution of Wireless Data Services: IS-95 to cdma2000", IEEE Communications Magazine, October 1998, pp. 140-149.

[Koh95]   R. Kohno, R. Meidan, L.B. Milstein, "Spread Spectrum Access Methods for Wireless Communications", IEEE Communications Magazine, January 1995, pp. 58-67.

[Kuc91]   A. Kucar, "Mobile Radio: an Overview", IEEE Communications Magazine, November 1991, pp. 72-85.

[Las97]   J.D. Laster, J.H. Reed, "Interference Rejection in Digital Wireless Communications", IEEE Signal Processing Magazine, May 1997, pp. 37-62.

[Lee93]   W.C.Y. Lee, Mobile Communications Design Fundamentals, New York: Wiley, 1993.

[lucen]   http://www.lucent.com

[Mac99]   D.J.C. MacKay, "Good Error Correcting Codes Based on Very Sparse Matrices", IEEE Trans on Inf. Theory, Vol. 45, No. 2, pp 399-431, 1999.

[Mad94]   U. Madhow, M.L. Honig, "MMSE Interference Suppression dor Direct-Sequence Spread-Spectrum CDMA", IEEE Transactions on Communications, Vol. 42, No. 12, December 1994, pp. 3178-3188.

[Mat02]   P. Mate, "Redefining mobile communications", Electronics Design Chain, Vol. 1, winter 2002, pp. 28-32.

[mathw]   http://www.mathworks.com

[McC73]   J.H. McClellan, T.W. Parks, L.R. Rabiner, "A Computer Program for Designing Optimum FIR Linear Phase Digital Filters", IEEE Transactions on Audio and Electroacoustics, December 1973, pp. 506-526.

[Men97]   U. Mengali, N.A. D'Andrea, Synchronization Techniques for Digital Receivers, Plenum Press, 1997.

[mento]   http://www.mentor.com

[mips]   http://www.mips.com

[Moe91]   T. Jesupret, M. Moeneclaey, G. Ascheid, "Digital Demodulator Synchronization - Performance Analysis", ESA/ESTEC Contract No. 8437/89/NL/RE, June 1991.

[Moe94]   M. Moeneclaey, "Overview of Digital Algorithms for Carrier Frequency Synchronization", In the Proceedings of the 4th International Workshop on DSP Techniques Applied to Space Communications, London, United Kingdom, September 1994.

[Mos96]   S. Moshavi, "Multi-User detection for DS-CDMA Communications", IEEE Communications Magazine, October 1996, pp. 124-136.

[motor]   http://www.motorola.com

[MUS01]   "MUSIC Receiver (Multi-User & Interference Cancellation)", ESA/ESTEC Contract 13095/98/NL/SB, Final Report, Noordwijk, The Netherlands, Dec. 2001.

[nalla]   http://www.nallatech.com

[Nat89]   F. Natali, "AFC Tracking Algorithms", IEEE Transactions on Communications, Vol. 37, No. 8, August 1989.

[Nee99]   R. van Nee, G. Awater, M. Morikura, H. Takanashi, M. Webster, K.W.Halford, "New High-Rate Wireless LAN Standards", IEEE Communications Magazine, December 1999, pp.82-88.

[New94]   P. Newson, M. Heath, "The Capacity of a Spread Spectrum CDMA System for Cellular Mobile Radio with Consideration of System Imperfections", IEEE Journal on Selected Areas in Communications, Vol. 12, No. 4, May 1994, pp. 673-684.

[Oja98]   T. Ojanperä, R. Prasad, "An Overview of Air Interface Multiple Access for IMT-2000/UMTS"; IEEE Communications Magazine, September 1998, pp. 82-95.

[Opp75]   A.V. Oppenheim, R.W. Shafer, Digital Signal Processing. Englewood Cliffs, NJ, Prentice Hall, 1975

[Pad95]   J.E. Padgett, C.G. Günther, T. Hattori, "Overview of Wireless Personal Communciations", IEEE Personal Communications Magazine, January 1995, pp. 28-41.

[Pav02]   M. Pavesi, "FlexBench Tool Suite Relies on Xilinx Silicon and Software", Xcell Journal, Issue 42, Spring 2002, pp. 64-67.

[Pet72]   W.W. Peterson, E.J. Weldon Jr., Error-Correcting Codes, 2nd ed., Cambridge, MA: MIT Press, 1972.

[Pic82]   R.L. Pickholtz, D.L. Schilling, L.B. Milstein, "Theory of Spread-Spectrum Communications-A Tutorial", IEEE Transactions on Communications, Vol. 30, Mat 1982, pp. 855-884.

[Pra98]   R. Prasad, T. Ojanperä, "An Overview of CDMA Evolution Toward Wideband CDMA", IEEE Communications Surveys, Vol. 1, No. 1, 4th Quarter 1998, pp. 2-29.

[Pro95]   J.G. Proakis, Digital Communications - 3rd edition, McGraw-Hill, New York, 1995.

[Rab00]   J.M. Rabaey, "Low-power silicon architectures for wireless communications", Proceedings of the IEEE ASP-DAC 2000, pp. 377-380, Yokohama, Japan, 25-28 January 2000.

[Rap91]   T.S. Rappaport, "The Wireless Revolution", IEEE Communications Magazine, November 1991, pp. 52-71.

[Rom00]   J. Romero-García, R. De Gaudenzi, F. Giannetti, M. Luise, "A Frequency Error resistant Blind CDMA Detector", IEEE Transactions on Communications, Vol. 48, No. 7, July 2000, pp. 1070-1076.

[Rom97]   J. Romero-Garcia, R. De Gaudenzi, F. Giannetti, M. Luise, "A Frequency Error Resistant Blind CDMA Detector", IEEE Transactions on Communications, Vol. 48, No. 7, July 2000, pp. 1070-1076

[Sam88]   H. Samueli, "The Design of Multiplierless FIR Filters for Compensating D/A Converter Frequency Response Distortion", IEEE Transactions on Circuits and Systems, August 1988, pp. 1064-1066.

[Sar80]   D.P. Sarwate, M.B. Pursley, "Crosscorrelation Properties of Pseudorandom and Related Sequences", Proceedings of the IEEE, Vol. 68, No. 5, May 1980, pp. 593-619.

[Sas98]   A. Sasaki, M. Yabusaki, S. Inada, "The Current Situation of IMT-2000 Standardization Activities in Japan", IEEE Communications Magazine, September 1998, pp. 145-153.

[Sch82]   R.A. Scholtz, "The Origins of Spread-Spectrum Communications", IEEE Transactions on Communications, Vol. 30, May 1982, pp. 822-854.

[Sei91]   S.Y. Seidel, T.S. Rappaport, S. Jain, M.L. Lord, R. Singh, "Path Loss, Scattering, and Multipath Delay Statistics in Four European Cities for Digital Cellular and Microcellular Radiotelephone", IEEE Transactions on Vehicular Technology, Vol. 40, No. 4, November 1991, pp. 721-730.

[Sim85]   M.K. Simon, J.K. Omura, R.A. Scholtz, B.K. Levitt, Spread Spectrum Communications, Rockville, MD: Computer Science Press, 1985.

[Smi97]   M. Smith, Application Specific Integrated Circuits, Addison-Wesley, 1997

[stm]   http://www.st.com

[Syn98]   R. De Gaudenzi, F. Giannetti, M. Luise, "Signal Synchronization for Direct-Sequence Code-Division Multiple Access Radio Modems", European Transactions on Telecommunications, Vol. 9, No. 1, January/February 1998, pp. 73-89.

[synop]   http://www.synopsys.com

[syste]   http://www.systemc.org

[tensi]   http://www.tensilica.com

[ti]   http://www.ti.com

[Vem96]   S. Vembu, A.J. Viterbi, "Two Different Philosophies in CDMA – A Comparison", In the Proceedings of the IEEE 46th Vehicular Technology Conference VTC'96, Atlanta, April 28-May 1, 1996, pp. 869-873

[Ver86]   S. Verdù, "Minimum Probability of Error for Asynchronous Gaussian Multiple Access Channels"; IEEE Transactions on Information Theory, Vol. 32, No. 1, January 1986, pp. 85-96.

[Ver97]   S. Verdù, "Demodulation in the Presence of Multiuser Interference: Progress and Misconceptions", in Intelligent Methods in Signal Processing and Communications, D. Docampo, A. Figueiras-Vidal, F. Perez-Gonzales, Eds. pp. 15-44, Birkhauser, Boston, 1997.

[Vit95]   A. Viterbi, S. Vembu, "Theoretical and Practical Limits of Wireless Multiple Access Systems", in the Proceedings of the URSI Commsphere, Eilat, Israel, April 1995, pp. 33-43.

[vsi]   http://www.vsi.org

Index

Numerics

1G system *See also* first generation system

2G system *See also* second generation system

3G system *See also* third generation system

4G system *See also* fourth generation system

A

AAF *See also* anti-alias filter

AC user location *See also* average case user location

A-CDMA *See also* asynchronous code division multiple access

AceS 9

acquisition
  average time 108
  parallel 105
  serial 105
  total time 107

ad hoc wireless network 1, 257

AD-AFC *See also* angle doubling automatic frequency control

adaptive detector 71

adaptive interference mitigation architecture 197

ADC *See also* analog to digital conversion

additive white gaussian noise 34

additive white gaussian noise generation 223

advanced mobile phone system 2

advanced mobile satellite task force 10

AFCU *See also* automatic carrier frequency control unit

AGC *See also* automatic gain control

algorithm
  Parks-McClellan 101
  Viterbi 66

alias profile 98

AMPS *See also* advanced mobile phone system

analog signal conditioning unit 82

analog to digital conversion 40

analog to digital converter 15

angle doubling automatic frequency control 119

antenna reflector 11

anti-alias filter 83

anti-image filtering 224

application specific integrated circuit 15, 255, 161

arbitrary waveform generator 224

ASCU *See also* analog signal conditioning unit

ASIC *See also* application specific integrated circuit

ASIC
  back-end design flow 216
  front-end design flow 211

Printed by Publishers' Graphics LLC